An Anthropology of Puzzles

An Anthropology of Puzzles

The Role of Puzzles in the Origins and Evolution of Mind and Culture

Marcel Danesi

Routledge
Taylor & Francis Group
LONDON AND NEW YORK

First published 2019 by Bloomsbury Academic

2 Park Square, Milton Park, Abingdon, Oxon, OX14 4RN
605 Third Avenue, New York, NY 10017

Routledge is an imprint of the Taylor & Francis Group, an informa business

First issued in paperback 2020

Cover image: PictureLake/Getty

Library of Congress Cataloging-in-Publication Data
Names: Danesi, Marcel, 1946– author.
Title: An anthropology of puzzles : the role of puzzles in the origins and evolution of mind and culture / Marcel Danesi.
Description: London, UK ; New York : Bloomsbury Publishing, Plc, 2018. |
Includes bibliographical references and index.
Identifiers: LCCN 2018030442| ISBN 9781350089853 (hb)
Subjects: LCSH: Puzzles—Social aspects. | Puzzles—Psychological aspects. | Cognition. | Human evolution. | Brain—Evolution.
Classification: LCC GV1493 .D33197 2018 | DDC 793.73—dc23
LC record available at https://lccn.loc.gov/2018030442

Typeset by RefineCatch Limited, Bungay, Suffolk

ISBN 13: **978-1-350-08985-3 (hbk)**
ISBN 13:978-0-367-73226-4 (pbk)

Contents

Figures

Tables

Preface

A good puzzle, like virtue, is its own reward.

Henry E. Dudeney (1857–1930)

Puzzles are everywhere—in magazines, in newspaper sections, and on countless websites. Why do we have this penchant for puzzles, no matter what our IQ or interests and talents? Is this desire nothing more than the behavior of a bored populace consuming puzzles to gain temporary recreation? As will be argued in this book, there is more to puzzles than recreation. They have been around since the dawn of history and have guided the evolution of large sections of that history, from discovery in mathematics to disquisitions on the nature of logic. Indeed, the human brain may itself be a "ludic organ," that is, an organ that impels us to gain knowledge through playful formats and artifacts of all kinds. This may appear to be somewhat of a grandiose claim; but it is not if one looks at human history from the angle of the many puzzle genres that have been a part of it since the beginning. Mathematical discovery is often a direct consequence of puzzles such as Alcuin's River-Crossing Puzzle or Euler's Königsberg Bridges Puzzle. The former contains the blueprint for critical path theory and combinatorics, and the latter for graph theory and topology. One of the oldest school textbooks of human civilization, the Egyptian *Ahmes Papyrus*, which dates back to before 1650 BCE, turns out to be essentially a collection of what we would call today mathematical brainteasers. In a phrase, puzzles have existed across time and across societies, revealing how human intelligence is embedded in a playful imagination.

This book argues that puzzles are as important as are the other artifacts that anthropologists and archeologists study and utilize to reconstruct and understand the origins of mind and culture. The term coined in 1974 by the American puzzlist, Will Shortz, to identify such study as a significant one for anthropology and psychology is *enigmatology*. This book can be considered to be a general overview of what enigmatology entails. People from time immemorial have always been fascinated by puzzles, becoming engaged with them for no apparent reward other than the simple satisfaction of solving them. Our instinctual love of puzzles may well be the product of neural mechanisms that produce our particular form of intelligence. I have written this book both to examine this possibility and to convey the intriguing psychological and anthropological story that puzzles tell us about ourselves. I hope to impart to the reader the sense that there is much more to puzzles than recreation or delectation; they are unique revelations of how we think and how we act upon the world.

Marcel Danesi
Fields Institute for Research in Mathematical Sciences,
University of Toronto, 2018

Acknowledgments

First and foremost, I would like to thank Miriam Cantwell and Lucy Carroll, my editors at Bloomsbury, for encouraging me to realize this book and for assisting me through all the phases of its editing and production. I cannot thank them enough for their support and enthusiasm. I am also deeply grateful to Victoria College of the University of Toronto for granting me the privilege of teaching a course on the history and meaning of puzzles for many years. This has allowed me to learn a great deal about the subject matter from the enthusiastic students I have taught. I have learned more from them than they have from me. I also want to express a debt of gratitude to Stacy Costa, the teaching assistant for the course, who has helped me enormously through the years. Finally, a heartfelt thanks goes out to my family for all the patience they have had with me over the years. I dedicate this book to my late father, Danilo, a simple and kind soul who inspired generosity and benevolence in all those around him.

1

Puzzles in Mind and History

A good puzzle, it's a fair thing. Nobody is lying. It's very clear, and the problem depends just on you.

Ernö Rubik (b. 1944)

One of the oldest riddles of recorded history is the Riddle of the Sphinx. In one widely told version, the Sphinx was a mammoth creature, half human, half animal, who had enslaved the city of Thebes, stopping all those who dared enter or leave it. The Sphinx posed a riddle to all foolish visitors as a mortal challenge. Those who were incapable of answering it paid for their ineptitude with their lives at the hands of the monster. However, if someone were ever able to come up with the seemingly intractable, yet simple, answer, the Sphinx vowed to destroy itself:

What creature ambles on four at dawn, two at midday, and three at twilight?

According to legend, it was Oedipus who solved the riddle by answering "*man*," who is the only creature on earth who crawls on all fours in infancy (the dawn of life), walks upright on two legs as a grown-up (the midday of life), and ends up walking on three, with the help of a walking stick, in old age (the twilight of life). Upon hearing the correct answer, the Sphinx jumped from its perch to a rock outside the city, becoming a lifeless statue. For ridding them of this terrible beast, the Thebans made Oedipus their king.

Why is this riddle so fascinating, still holding appeal millennia later? For one thing, it is an ingenious formulation of something that might otherwise escape attention—the phases of human life are like the phases of a day, implying that we cannot escape our mortal destiny in the same ineluctable way that the twilight inevitably comes. The riddle is a story within a story. The main one is the Oedipus legend, which recounts a self-fulfilling prophecy. Oedipus had been left to die on a mountain by his father Laius, who had been warned by an oracle that he would be killed by his own son. The infant Oedipus was saved by a shepherd. After growing up, and learning about his origins, the young Oedipus traveled to Thebes in search of the truth. After solving the Sphinx's riddle, the Thebans made him the successor to their murdered king, who, unbeknownst to Oedipus, was his father. As the new king, Oedipus married Laius's widow, Jocasta. Several years later, a plague struck the city. An oracle announced that the scourge would come to an end only after Laius's murderer had been driven from Thebes. Oedipus investigated the murder and soon realized that Laius was the man he had killed on the road to Thebes. To his horror, he also discovered that Laius was his father and Jocasta his mother. Grief-stricken and desperate, Oedipus blinded himself, and Jocasta hanged herself. Oedipus was banished from Thebes, dying in unendurable woe at Colonus. The subtext in this legend is a significant one for the purposes of this book—riddles may be warnings about the realities of the human condition.

Various versions of the Sphinx's riddle exist. The one above is paraphrased from the play *Oedipus Rex* by Sophocles. Whatever its version, it is evidence that, since the dawn of history, people have devised riddles to understand themselves and the world around them. These might therefore reveal how sentient reflective thought emerged in our species, constituting miniature models of how we grasp things. This chapter provides an overview of the origin, history, and connection of puzzles to human thought and culture. The purpose is to argue in an initial way that puzzles arise from a deep-seated need to ask questions about existence. They do so in their own miniature way, constituting small-scale versions of the larger-scale questions of philosophy and science.

Riddles, puzzles, and games

There is no historical era or culture without riddles. They constitute a universal speech art that cuts across all languages. This is perhaps why riddles lose almost nothing in translation, tapping into common themes of human concern, from mortality to the meanings of things. The topic of riddles will be explored in more detail in the next chapter. Suffice it to say here that we continue to cherish and appreciate them, no matter who created them or when they were devised. Like works of visual art, they never lose their aura, being passed on from generation to generation intact.

The English word *puzzle*, as used today, encompasses everything from riddles and crosswords to Sudoku, optical illusions, and brainteasers in logic and mathematics. The word was coined near the end of the sixteenth century, and applied a little later to describe the jigsaw puzzle. As a generic categorical term, *puzzle* is a convenient one for classifying diverse enigmatological artifacts as singular psychological and anthropological phenomena.

Labeling some ancient riddle or puzzle, therefore, is a retrospective form of reference. The concept of *problem* came out of ancient geometry, where it was used to refer to a proposition in which some shape or figure had to be constructed. From this, the word was extended to cover any mathematical question that required a specific kind of answer and a strategy to do so. Many of the ancient mathematical problems were actually puzzles, as we would name them today. The main difference is that the intent of a problem is to produce a specific and recognizable answer; the intent of a puzzle, on the other hand, is to hide the answer, presenting information that appears to be incomplete in some way. Both problems and puzzles are Q & A (Question and Answer) structures. So, the difference between the two can be shown diagrammatically as follows:

Problem
Q → A (the question leads directly to an answer)

Puzzle
Q → (A) (the answer to the question is not immediately obvious)

Riddles have the same structure and it is for this reason that they can be called puzzles, even though there are some key differences between riddles and other puzzle inventions (as will be discussed in the next chapter). As the Oedipus story indicates, riddles were often connected to ominous destiny, unlike problems in geometry. But riddles also had a recreational social function. The Biblical kings Solomon and Hiram, for example, organized riddle contests simply for the pleasure of outwitting each other. The Greeks included riddles at banquets, as we might do today at social gatherings, for entertainment reasons. The Romans made riddles a central feature of the Saturnalia, a feast that they celebrated over the winter solstice. So, riddles were conceived both as part of portentous myth, used by the Greek oracles to cast fortunes, and as part of recreation. This dual function extended throughout the medieval and Renaissance periods. Only by the eighteenth century, did they lose their divinatory value, becoming mainly forms of mind-play, included as regular features by newspapers and periodicals. It was then that famous personages started creating riddles as part of an ever-expanding leisure culture and as part of a new literary genre. Benjamin Franklin, for example, composed riddles under the pen name of Richard Saunders for inclusion in his *Poor Richard's Almanack* (first published in 1732). The puzzle section was a factor in the almanac's unexpected success. In France, no less a literary figure than the satirist Voltaire would regularly compose riddles for pure enjoyment and to challenge or taunt his friends and enemies.

The Greeks saw riddles as manifestations of *mythos*—a form of thinking based on beliefs rather than on logical argumentation. They used the term *lógos* to describe the latter. Socrates believed that *lógos* was innate in all human beings, teaching that everyone had full knowledge of truth within them, and that this *a priori* knowledge could be accessed through conscious reflection. In the *Meno*, a Socratic dialogue written by Plato (2006, originally *c.* 380 BCE), Socrates leads an untutored slave to grasp a complicated geometrical problem by getting him to reflect upon the truths hidden within him through a series of questions designed to elicit specific answers. From this Q & A mode of dialogue, the concept of *dialectic* investigation crystallized, as the art of investigating and discussing the truth of ideas. The notion of *mythos* can be rephrased as "imagination" and *lógos* as "reasoning." As will be argued throughout this book, puzzles involve both modes of thought, to different and varying degrees. If we are given the circumference of a circle and asked to determine its radius, it is easy to figure it out if we know the formula C (circumference) = πr^2 (with r = radius). This is a simple problem based on using previous knowledge applied directly to a given situation—the solver can thus take a shortcut to *lógos*. Now, by contrast, consider the following, which at first glance would seem to suggest a similar kind of problem that can be solved just as directly. It was devised by Martin Gardner (1994: 14) and is shown in Figure 1.1.

Given the dimensions of the radius OD (6 + 4 = 10), can you calculate the length of the diagonal AB in rectangle AOBC?

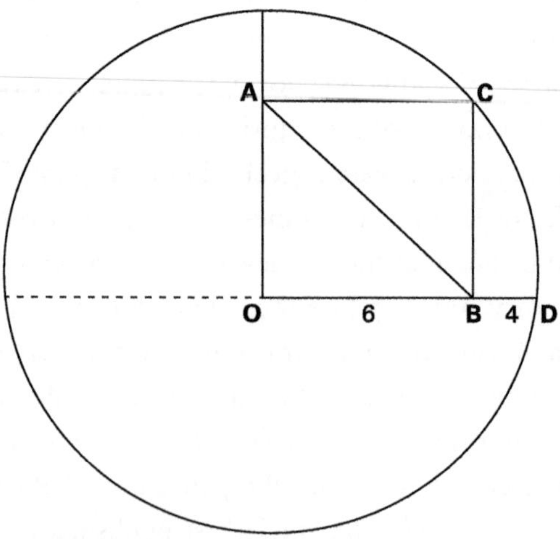

Figure 1.1 Gardner's diagonal puzzle.

As it turns out, it is impossible to solve it in an analogously straightforward fashion. So, let's consider what is known about circles and rectangles from a different perspective. As a hunch, let's use the information that the diagonals of a rectangle are equal to each other in length. This suggests drawing the other diagonal (OC) of the rectangle AOBC (Figure 1.2).

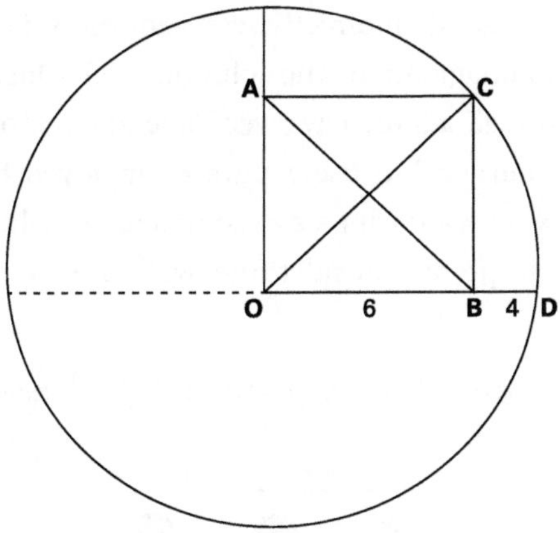

Figure 1.2 Solution to Gardner's diagonal puzzle.

By doing this, we can now see that diagonal OC is also a radius of the circle. We know that radii are equal from an established theorem. Line OBD is a radius and is equal to 10 (6 + 4), as shown. Since line OC is also a radius it is thus equal to 10. From this we conclude that the other diagonal, AB, is also equal to 10. The solution now appears almost magically, rather than routinely—hence the use of the expression "Aha" to characterize the effect a solution such as this one might have on us. The solution thus starts out as imaginative thinking (playing a hunch) and then ends with reasoning (carrying through on this hunch). It shows, in other words, a flow from *mythos* to *lógos*, not a separation of the two.

Most puzzles, are characterized by this flow, to varying degrees, of course. It is relevant to note here that imaginative thinking of this kind originates in the right hemisphere of the brain (the dreaming hemisphere)—a fact that is especially useful in contemplating the possible relation of puzzles to human consciousness. Solving a puzzle is very much like a dream where the parts cohere spontaneously into wholes by themselves. There is some archeological evidence that the difference between waking (consciousness) and dreaming (unconsciousness) states in antiquity was often blurred (Jaynes 1976). Ideas were believed to come in dreams (the unconscious mind), and they likely still do, as well-known episodes from mathematics strongly suggest. The famous

Indian mathematician, Srinivasa Ramanujan, claimed that his family goddess would come to him in his dreams and present him with complex mathematical equations, providing crucial insights that allowed him to complete his efforts. Similarly, Henri Poincaré (1902) observed that, typically, it was only after a period of sleep marked by dreaming that his imagination developed the relevant insights into the intractable mathematical conundrums that he tried to crack. As these anecdotes suggest, indirectly, consciousness is to reasoning what unconsciousness is to imagination. The solution to Gardner's puzzle involved a blend of the two. Through a hunch, we were able to envision a solution that lay "hidden behind the curtain," to use a phrase employed by Poincaré. At that point, reasoning took over, allowing us to complete the solution.

As another case-in-point, consider the well-known Nine-Dot Puzzle in Figure 1.3.

Without your pencil leaving the paper, can you draw four straight lines through the following nine dots?

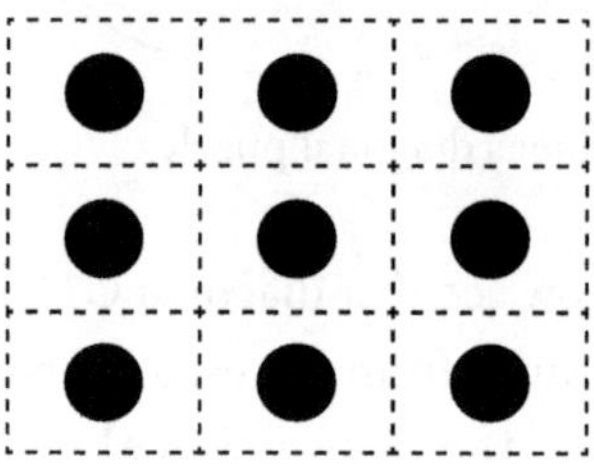

Figure 1.3 The Nine-Dot Puzzle.

Those unfamiliar with this puzzle tend to tackle it by joining up the dots as if they were located on the perimeter (boundary) of an imaginary square or flattened box (Figure 1.4).

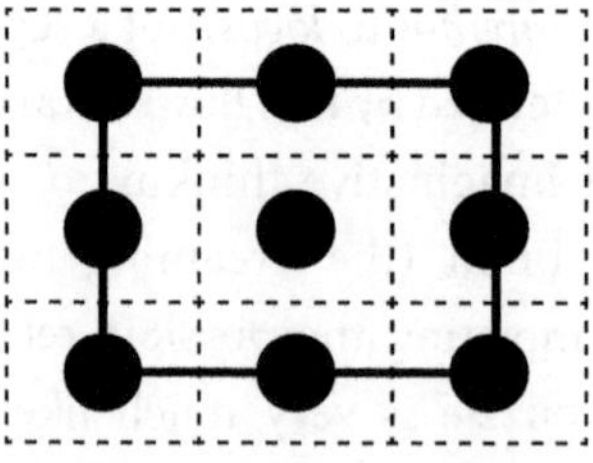

Figure 1.4 An incorrect solution to the Nine-Dot Puzzle.

But this reading of the puzzle does not yield a solution, no matter how many times one tries to draw four straight lines without lifting the pencil. A dot is always "left over." At this point, an imaginative hunch is needed, as with Gardner's puzzle: "What would happen if the four lines were to be extended beyond the apparent box structure of the dots?" That hunch turns out, in fact, to be the relevant Aha insight. One solution is as shown in Figure 1.5.

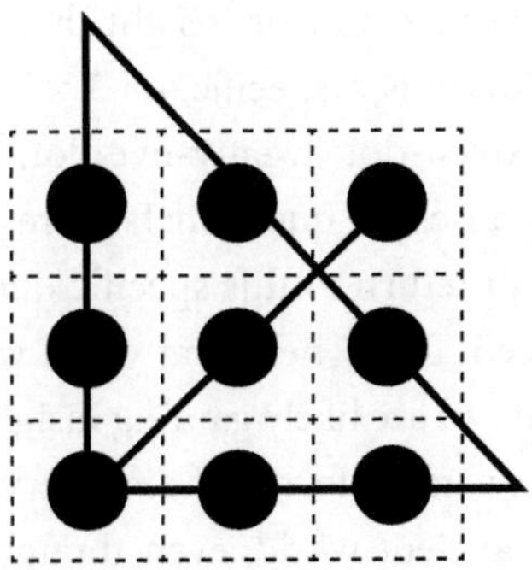

Figure 1.5 A solution to the Nine-Dot Puzzle.

Incidentally, this puzzle is the likely source of the common expression "thinking outside the box." The reason for this is self-explanatory—envisioning a solution based on perceiving the dots as forming a box-like figure will lead nowhere; the solution requires us to think outside that figure. The first appearance of a dot-joining puzzle is probably in the 1914 posthumous edition of American puzzlist Sam Loyd's *Cyclopedia of 5000 Puzzles, Tricks and Conundrums with Answers*. But the principle it embodies is likely older, as Martin Gardner indicates in his 1960 edition of Loyd's work (*The Mathematical Puzzles of Sam Loyd*). The puzzle is now commonly used in psychology to study lateral thinking (Kershaw and Ohlsson 2004). It requires solvers to literally look beyond figure-ground relations that present themselves on a piece of paper or screen. The term "lateral thinking" was proposed by psychologist Edward de Bono (1970) to refer to the kind of Aha thinking process involved in solving the Nine-Dot Puzzle. Incidentally, there is a clever version of this puzzle in Figure 1.6.

Can you connect all the dots with only three lines?

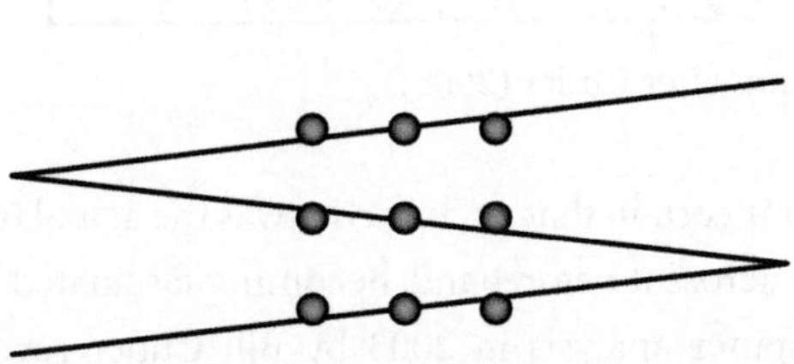

Figure 1.6 An ingenious version of the Nine-Dot Puzzle.

Notice that in this case the lines barely touch some of the dots—the puzzle did not ask us to go through them, thus adding a layer of ingenuity to it that makes it much more challenging. Now, a corollary about puzzles that will constitute a major theme of this book is that, typically, a particular solution is a snapshot of some inherent principle or insight that can be abstracted from it. The original Nine-Dot Puzzle is a specific, or 3 × 3 version, of a Dot-Joining Puzzle. There are also sixteen-dot, twenty-five-dot, and increasingly complex Dot-Joining Puzzles. This raises the question: Is there an abstract structure to the different solutions? We will return to this specific question vis-à-vis this puzzle in the final chapter. Suffice it to say here that when we start thinking abstractly beyond the original puzzle, we are likely gaining insights into the ludic mind and how it informs and guides various forms of discovery.

Puzzles existed in the ancient world, even though they were not named as such. However, the term *game* was used broadly. For instance, Archimedes' *loculus* was devised as a kind of ingenious game with mathematical implications. It had fourteen geometrical shapes, some of which were identical. The original objective was to scramble the shapes and then rearrange them so that they would form a square figure. The loculus can be partitioned as shown in Figure 1.7 (see Cutler 2003)—note that the two "6s" and two "7s" are labeled this way because they are identical.

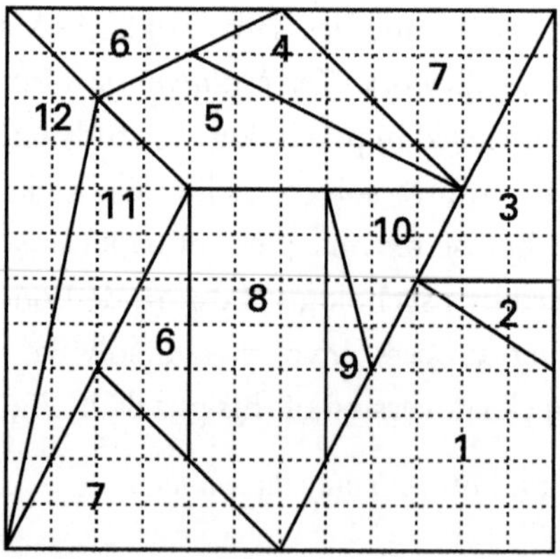

Figure 1.7 The *loculus* after Cutler (2003).

It is not known for certain that Archimedes was the actual inventor of the game; he may have come across it beforehand, becoming fascinated by its mathematical properties. A computer analysis in 2003 by Bill Cutler revealed that there are 536 distinct ways to arrange the pieces. Now, what is particularly interesting about this game is that its structure crops up subsequently in different formats and guises. For example, there are many parallels between the Archimedean *loculus* and modern-day tangram puzzles, which involve rearranging and combining seven flat geometrical shapes, called *tans*. In some versions of the Archimedean game, the shapes are used to form different figures (human shapes, animal shapes, object shapes, and so on)—an objective that applies to tangrams, which will be discussed in due course. For the sake of historical accuracy, it should be mentioned that the *loculus* was first discussed in a treatise attributed to Archimedes called the *Ostomachion*—a work that has survived in an Arabic version and in a palimpsest of the original Greek text going back to the Byzantine era (Darling 2004: 88).

Puzzles like the Nine-Dot one present a situation that seems to defy a solution. Games like the *loculus*, on the other hand, ask us to do something in a specific way (for example, rearrange pieces into something recognizable). In games, one can try out different arrangements by a kind of "brute force" method, whereby various arrangements are envisioned in the hope of eventually getting the correct one(s). Brute-force thinking rarely characterizes puzzles (Palmer and Rodgers 1983). This does not mean that games do not involve highly imaginative thinking. Consider the so-called Knight's Tour, which asks the following (Conrad et al. 1994):

Place a knight on the chessboard so that it visits every square once and only once.

Like the *loculus*, there are many solutions, with the earliest one dating back to the ninth century in the *Kavyalankara*, a Sanskrit work on the nature of poetry. Below is the "Knight's Graph," which shows all possible paths for the tour—the numbers indicate the possible moves that can be made from that position (see Figure 1.8 overleaf).

The Knight's Tour clearly involves a form of brute-force thinking for reaching the end-states (all possible tours). Nevertheless, as one works out the details of the various tours there is more than trial-and-error involved. In practice, the difference between puzzles and games is ignored by mathematicians, because the solutions to games, like the solutions to puzzles, bear various theoretical implications (as we shall see). There are four main types of games that are relevant to logical and mathematical analysis (Dalgety and Hordern 1999): movement and arrangement games, played by manipulating objects such as sticks, coins, or counters with the hands; mechanical and assembly games, played by assembling pieces to make shapes; board games, played on a game board; and card and dice games. These will be discussed subsequently.

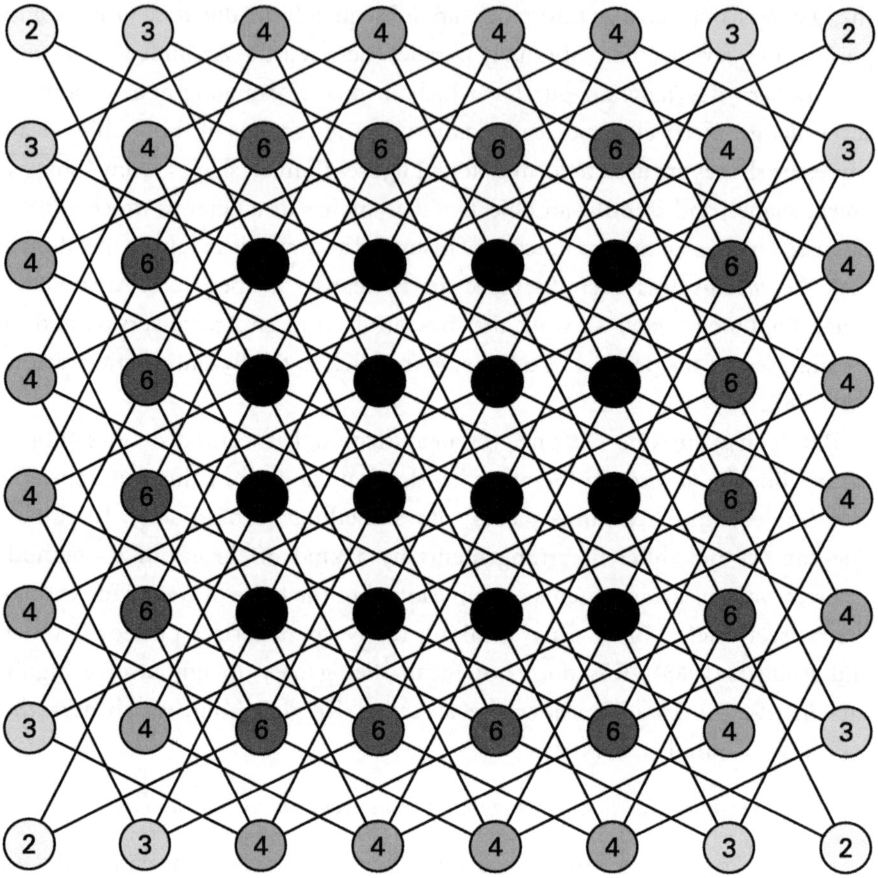

Figure 1.8 The Knight's Graph (from Wikimedia Commons).

One way to differentiate between the two is as open or closed enigmatic artifacts (Danesi 2018). The latter do not hide the answer, but give it at the start as an "end-state" that must be reached. The challenge inheres in *how* to get to the end-state. Games such as those in Figure 1.8 are closed puzzle texts, whereby we are given a set of rules or a set of conditions of how to reach the end-state. Open puzzles, on the other hand, do not involve end-states or rules. The answer in open puzzles is never obvious; the challenge is figuring out what the answer is, not how to get to it.

Puzzles and games put the mind's resources on display, including imagination, reasoning, memory (of relevant information), trial-and-error, inference, association, analogy, logic, and so on. It is impossible to predict beforehand which of these modes are deployed in tandem. But that unpredictability is precisely what makes puzzles so appealing and challenging. The "fun" is in reaching the Aha insight or the end-state. As Scott Kim (2016) aptly notes: "A puzzle is fun, and it has a right answer." While the specific characteristics of puzzles and games might vary in detail from culture to culture, their ludic nature and artful intent do not. Moreover, we rarely resist the challenge they pose. As Helene Hovanec (1978: 10) puts it, puzzles are irresistible because they "simultaneously conceal the answers yet cry out to be solved," piquing the ingenuity of solvers "against that of the constructors."

The study of the anthropological meanings of puzzles, riddles, and games can be assigned the general rubric of *enigmatology*, as mentioned in the Preface. Strangely, the term has rarely caught on in academia, despite the obvious implications that it has for various fields. Only in mathematics (as we shall see subsequently) has enigmatology—known more specifically as *recreational mathematics*—become an autonomous branch of the discipline. The reason for this may well be that, as Poincaré pointed out (above), the kind of thinking involved in puzzle-solving is a powerful one that often leads to unexpected discoveries. Throughout the history of mathematics, in fact, specific puzzles have led to general insights, which, in turn, have formed the basis of new branches. Zeno's paradoxes of motion, for example, led to the invention of the calculus, Alcuin's River-Crossing Puzzles prefigured modern-day critical path theory, Euler's Königsberg's Bridges Puzzle inspired graph theory and topology, and the list could go on and on. As Kasner and Newman (1940: 156) observed in their classic book on the mathematical imagination, the "theory of equations, of probability, the infinitesimal calculus, the theory of point sets, of topology, all have grown out of problems first expressed in puzzle form."

Historical sketch

There is no single history of puzzles, probably because there are so many genres and traditions scattered throughout the world that it is virtually impossible to tie them together into a unified chronological paradigm. It is more practicable to write separate histories of individual genres—riddles, logical conundrums, and so on (Singmaster 1996). In subsequent chapters, various historical annotations will be interspersed into the relevant discussions. Only a patchwork of events can be weaved together here into a very selective sketch that will serve primarily as a background to the overall objectives of this book. As Hovanec (1978: 10) cogently writes:

> Tracing the historical roots of puzzles is as revealing as constructing one's family tree. In investigating family lineage the blood relationships between generations yesterday and today are not always readily apparent; sometimes in-depth examination is necessary before the particular elements connecting a family's heritage can be recognized. Yet the ties that bind definitely do exist. So it is with puzzles. Most of today's puzzles differ dramatically from their predecessors; yet like families, the branches joining the past and present are visible, albeit faint. The roots are there and exploring them can help us appreciate the unique heritage of today's puzzles and the refinements which have occurred over the centuries.

Two bones were discovered by archeologists near Lake Edward in modern-day Zaire, dating back 11,000 years (Olivastro 1993: 5–11). They were made by the ancestors of the Ishango. No one knows for certain what their purpose was; but with our modern eyes they look suspiciously like a game in which the notches likely represented numbers. If the artifact was indeed a game, then it is clear evidence that mind-play coincides with the emergence of early human cultures. One of first documented collections of mathematical puzzles dates back to 1650 BCE in Egypt. Eighteen and a half feet long and thirteen inches wide, the text is referred to either as the *Ahmes Papyrus*, after the Egyptian scribe Ahmes, who copied it (making it even older in time), or the *Rhind Papyrus*, after the Scottish lawyer and antiquarian, A. Henry Rhind, who purchased it in 1858 while traveling in Egypt. The papyrus, Rhind discovered, had been found a few years earlier in the ruins of a small building in Thebes in Upper Egypt. We shall return to this papyrus subsequently. Suffice it to say here that it was composed about the same time as the Riddle of the Sphinx and other riddles, such as those found in the Bible and Sanskrit manuscripts. It also dovetails with the first labyrinths—ancient buildings designed as architectural puzzles. One of these, constructed by Pharaoh Amenemhet III in Egypt, dates back to the nineteenth century BCE as described by the Greek historian Herodotus (440 BCE, *The Histories*, Book II: 148).

One of the most famous labyrinths of antiquity was the prison built on the island of Crete, which may, however, have existed only in myth. The story goes that Androgeus, the son of King Minos of Crete, was brutally murdered by some Athenians. In fatherly despair and anger, Minos had a prison built to exact his revenge, which will be discussed in Chapter 4. Archaeologists have discovered a palace located in the Cretan city of Knossos that may have been the site of the mythical labyrinth, because it has passageways like those described in the legendary accounts of the building.

Labyrinths were likely designed as intelligence tests, separating heroes from common folk. Modern-day mystery stories often utilize this latent mythology associated with labyrinths. Novels such as Stephen King's *The Shining* (1977), Umberto Eco's *Name of the Rose* (1983), among others, revisit the Cretan labyrinth and its many psychological implications in contemporary ways. Now, while the mystery surrounding labyrinths may have faded, they continue to be built as tests of wits, both recreational and serious. Maze structures with confusing intertwining passages are commonly found in amusement parks. These are designed to hold people in suspense as they try to find their way through them. Psychologists use mazes more seriously to test problem-solving skills in animals and humans alike. And toy mazes are popular games given to children today mainly because they are thought to sharpen thinking skills, while providing diversion at the same time.

Many ancient puzzles were conceived as models of mathematical notions or principles. Archimedes, for instance, contrived his famous *Cattle Problem* as a way to discuss and illustrate properties of large numbers. The problem asks us to compute the number of cattle in a herd, given a number of restrictions. Its details need not concern us here. It was discovered by mathematician Gotthold Ephraim Lessing in 1773 (see Lessing 1905). Below is a version of the problem:

> *There is a herd of cattle consisting of bulls and cows, one part of which is white, a second black, a third spotted, and a fourth brown. Among the bulls, the number of white is one half plus one third the number of the black greater than the brown; the number of the black, one quarter plus one fifth the number of the spotted greater than the brown; the number of the spotted, one sixth and one seventh the number of the white greater than the brown. Among the cows, the number of white is one third plus one quarter of the total black cattle; the number of the black, one quarter plus one fifth the total of the spotted cattle; the number of spotted, one fifth plus one sixth the total of the brown cattle; the number of the brown, one sixth plus one seventh the total of the white cattle. What was the composition of the herd in terms of white and black bulls and cows?*

The problem remained unsolved for a number of years, mainly because of the huge numbers and complex calculations involved. A solution was published in 1880 by A. Amthor and B. Krumbiegel. There are actually various results that can be obtained from the problem, producing numbers with 206545 digits (Williams,

German, and Zarnke 1965). Ilan Vardi (1998) developed simple formulas to generate the solutions to the cattle problem, revealing that below the seemingly complex computational structure there is a simple idea. This is, in fact, a salient characteristic of all the classic mathematical puzzles (as will be discussed). In his book, *The Sand Reckoner*, Archimedes (1897) deals more specifically with the dilemma of large computations via the following numerical conundrum: *What is the upper limit for the number of grains that fit into the universe?* Archimedes' answer of 10^{63} grains is intriguing, because it was more or less verified by physicist Arthur Stanley Eddington in 1938. Eddington posited the number of protons in the universe to be around 10^{80}. Since Archimedes' 10^{63} grains contain roughly 10^{80} protons, the two numbers are essentially equal. The list of ancient mathematicians who devised puzzles and games as abstract models of structure is extensive. Heron of Alexandria, who was known for mechanical inventions such as a rotary steam engine and a primitive surveying instrument, contrived ingenious puzzles, as concrete vehicles, for investigating square and cube roots. In his famous treatise, *Arithmetica* (1910, originally *c.* 250 CE), the Alexandrian mathematician Diophantus constructed puzzles specifically to illustrate methods of solving algebraic equations.

Collections of riddles and puzzles became rather common in the medieval period. One of the best known is the *Greek Anthology*, compiled by the Greek poet, grammarian, and mathematician Metrodorus (1916–1918, originally *c.* 500 CE). Among the poems and epigrams that he put into his collection, we find a chapter of mathematical puzzles, enigmas, and oracles (riddles). What we have here is, arguably, the first puzzle collection in the modern sense. A few centuries later, puzzledom's first masterpiece appears, the *Propositiones ad acuendos juvenes*, by the English scholar and ecclesiastic Alcuin of York around 800 CE (translated into English as "Problems to Sharpen the Young" by Hadley and Singmaster in 1992). The book consists of over fifty puzzles intended to train medieval youths in mathematics and logical thinking. It was Charlemagne who had invited Alcuin to teach his own children and himself. A number of Alcuin's puzzles have become classics of recreational mathematics. The most widely known one is the following:

> *A traveler comes to a riverbank with a wolf, a goat and a head of cabbage. There is a small rowing boat that he can use to get across, but it is only big enough to take himself and one other. If he leaves the goat and the wolf by themselves on either side, the wolf will eat the goat. If he leaves the cabbage with the goat alone, the goat will eat the cabbage. How can he get both animals and the cabbage safely across the river?*

This puzzle always seems to strike a resonant chord in people, no matter their background. It portrays a situation that involves decision making and planning. The crucial decision is the first one: "What would happen if . . .?" The answer is the key to solving the puzzle. Disaster would occur if the traveler started with either the cabbage or the wolf. In the former instance, the wolf would eat the goat; in the second, the goat would eat the cabbage. So, the traveler must start with the goat, leaving the wolf and goat safely behind. The rest of the reasoning is straightforward. The traveler goes over to the other bank, leaving the goat and coming back alone. He can then pick up either the cabbage or the wolf. Let's say he picks up the wolf. Once over on the other side, he drops off the wolf and goes back with the goat (for obvious reasons). When he gets back, he drops off the goat and picks up the cabbage. On the other side, he drops off the cabbage, leaving it safely with the wolf. He goes back alone, picks up the goat and when he reaches the other side, continues on with his journey. More will be said about this puzzle. The point here is that it illustrates a form of decision-making that goes on all the time but rarely recognized as harboring some distinctive abstract property.

Olivastro (1993: 129) suggests that Alcuin's collection may have been designed initially for the diversion of Charlemagne, that is, as a kind of recreational book. The evidence comes from a 799 CE letter from Alcuin to Charlemagne:

> I have sent to your excellency some figures of arithmetical subtlety for your amusement, on the blank part of the paper which you have sent us; in order that what offered itself to our eye naked may come back to you.

The letter does not contain the puzzles, and so it is not possible to truly decipher the intent of the letter. Whatever the case, Alcuin's puzzles were written with a literary flair, resembling in style the ancient fables of Aesop (Olivastro 1993: 129).

After Alcuin, puzzle-making recedes somewhat from the social scene, with only sporadic examples of riddles and games dispersed in various medieval texts. It resurfaces in the early thirteenth century with Leonardo Fibonacci's *Liber Abaci*, published in 1202 (Fibonacci 2002). The Rabbit Puzzle, found in the third section of the book, has become emblematic of recreational mathematics as a distinct branch. It will be examined subsequently; here, it is sufficient to paraphrase it and discuss it schematically:

> *A certain man put a pair of rabbits, male and female, in a very large cage. How many pairs of rabbits can be produced in that cage in a year if every month each pair produces a new pair which, from the second month of its existence on, also is productive?*

The solution involves envisioning a sequence of events connected in some regular way. There is one pair of rabbits in the cage at the start. At the end of the first month, there is still only one pair, for the puzzle states that a pair is productive only "from the second month of its existence on." It is during the second month, therefore, that the original pair will produce its first offspring pair. Thus, at the end of the second month, a total of two pairs, the original one and its first offspring pair, are in the cage. Now, during the third month, only the original pair generates another new pair. The first offspring pair must wait a month before it, too, becomes productive. So, at the end of the third month, there are three pairs in total in the cage—the initial pair, and the two offspring pairs that the original pair has thus far produced. If we keep tabs on the situation month by month, we can show the sequence of pairs that the cage successively contains as follows:

1, 1, 2, 3

The first digit represents the number of pairs in the cage at the start; the second, the number after one month; the third, the number after two months; and the fourth, the number after three months. During the fourth month, the original pair produces yet another pair. At that point in time the first offspring pair produces its own offspring pair. The second pair produced by the original rabbits has not started producing yet. Therefore, during that month, a total of two newborn pairs of rabbits are added to the cage. Altogether, at the end of the month there are the previous three pairs plus the two newborn ones, making a total of five pairs in the cage. This number can now be added to our sequence:

1, 1, 2, 3, 5

During the fifth month, the original pair produces yet another newborn pair; the first offspring pair (now fully productive) produces another pair of its own as well (its second); and now the second offspring pair produces its first pair. The other rabbit pairs in the cage have not started producing offspring yet. So, at the end of the fifth month, three newborn pairs have been added to the five pairs that were previously in the cage, making the total number of eight pairs in it. We can now add this number to our sequence:

1, 1, 2, 3, 5, 8

Continuing to reason in this way, it can be shown that after twelve months there are 233 pairs in the cage:

1, 1, 2, 3, 5, 8, 13, 21, 34, 55, 89, 144, 233

The salient (and initially hidden) characteristic of this sequence is recursion. Each number in it is the sum of the previous two: for example, 2 (the third number) = 1 + 1 (the sum of the previous two); 3 (the fourth number) = 1 + 2 (the sum of the previous two); and so on. This pattern can be generalized by using simple algebraic notation. If we let F_n stand for any "Fibonacci number," and F_{n-1} the number just before it, and F_{n-2} the number just before that, the recursion pattern inherent in the sequence can be shown as follows:

$$F_n = F_{n-1} + F_{n-2}$$

This means that the sequence can be extended ad infinitum:

$$\{1, 1, 2, 3, 5, 8, 13, 21, 34, 55, 89, 144, 233, 377, 610, 987, \ldots\}$$

Little did Fibonacci know how significant his sequence would become. Over the years, the properties of Fibonacci numbers have been studied extensively, resulting in a considerable specialized technical and popular literature. The recursive pattern was first detected and studied by the French mathematician Albert Girard in 1632. In 1753, the Scottish mathematician Robert Simson noted that, as the numbers increased in magnitude, the ratio between succeeding numbers approached the *golden ratio*, whose value is 0.618.... The list of discoveries, mathematical and otherwise, in the Fibonacci sequence is truly astounding—so much so that a journal, called *The Fibonacci Quarterly*, was established in 1963 to publish new discoveries, discussions, debates, and findings of the sequence and its derivatives. Why would the solution to an artificial puzzle hide so many patterns and ideas within it? There is, to the best of my knowledge, no definitive answer to this question. As mathematician Ian Stewart (2001: v) has elegantly put it, "simple puzzles could open up the hidden depths of the universe." As a model prefiguring contemporary computer science and theories of language, Fibonacci's puzzle harbored mysteriously within it a deep insight—namely that recursion might constitute the fabric of a slice of reality.

In the same century as the *Liber Abaci*, puzzles and games became a sort of craze, as evidenced by publications such as the *Book of Games* (Mohr 1993), commissioned by King Alfonso X of Castile and Léon (*c.* 1226–84), which contained clear instructions of how to play chess, checkers, and various card and board games. Chess had become so popular that it was adopted by mathematicians to explore ideas in ingenious ways. One of the most famous chess puzzles of that era is by Ibn Khallikan (*c.* 1256), an Islamic judge and author of a classic Arabic biographical dictionary. It is paraphrased below:

How many grains of wheat are needed on the last square of a 64-square chessboard if 1 grain is to be put on the first square of the board, 2 on the second, 4 on the third, 8 on the fourth, and so on in this fashion?

If one grain (= 2^0) is put on the first square, two grains (= 2^1) on the second, four on the third (= 2^2), eight on the fourth (= 2^3), and so on, it is obvious that 2^{63} grains will have to be placed on the sixty-fourth square. Khallikan's puzzle illustrates, in effect, the practical meaning of a geometric progression with general term 2^n. The term 2^{63} is the sixty-fourth one in the sequence and can be represented as 2^{n-1}. The value of 2^{63} is so large that it boggles the mind to imagine what kind of chessboard could hold so many grains, not to mention where so much wheat could be harvested. Khallikan's puzzle is somewhat reminiscent of Archimedes' sand-reckoner conundrum, which asks us to compute the number of grains of sand it would take to fill the universe. Versions of Khallikan's puzzle crop up in other areas, such as India, suggesting that it might harbor some embedded archetype—a theme to be explored throughout this book.

During the Renaissance, puzzle-making became regular as part of mathematical investigations, with scholars such as Robert Recorde, Niccolò Fontana Tartaglia, and Girolamo Cardano increasingly employing the puzzle format to study or model or explore mathematical concepts. By the early seventeenth century, puzzle compilations had become widespread throughout Europe, both as materials for delectation and as tools for illustrating and probing philosophical and mathematical ideas. In 1612, the French Jesuit poet and scholar Claude-Gaspar Bachet de Mézirac published a collection of puzzles, titled *Problèmes plaisans et délectables qui se font par les nombres* (Bachet 1984), that continues to stimulate interest to this day. Bachet put forward the first-ever classification of mathematical puzzles—weighing puzzles, number tricks, and so on—a format that has been adopted *grosso modo* by puzzlists ever since. Another widely known puzzle anthology of the same era was Henry van Etten's *Mathematical Recreations, or, A Collection of Sundrie Excellent Problemes out of Ancient and Modern Phylosophers Both Usefull and Recreative*, published in French in 1624 and then in English in 1633.

In the eighteenth century, mathematicians continued to create puzzles as a means of engaging creatively with their subject matter. With his Thirty-Six Officers Puzzle of 1779, Leonhard Euler aroused interest in a fledgling area of mathematics at the time, subsequently named combinatorics. The puzzle asks: *Is it possible to arrange 6 regiments consisting of 6 officers each of different rank, in a 6 × 6 square so that no rank or regiment will be repeated in any row or column?*

Such an arrangement turns out to be impossible, but the puzzle quickly became one of the cornerstones of combinatorics. Euler also invented a puzzle seemingly for the enjoyment of it—a puzzle form that prefigured Sudoku, called a Latin Square, or an arrangement of letters in which each one occurs n times in a square array of n^2 cell so that no letter appears twice in the same row or column. An example of a Latin Square is shown in Figure 1.9—in it, the letters A, B, C, D appear once and only once in each row and column.

A	B	D	C
B	C	A	D
C	D	B	A
D	A	C	B

Figure 1.9 A Latin Square.

Euler's best-known puzzle is, however, his Königsberg's Bridges Puzzle, which is really an ingenious treatise on graphs and networks. In the German town of Königsberg runs the Pregel River. In the river there are two islands, which in Euler's times were connected to the mainland and to each other by seven bridges. The residents of the town would often debate whether or not it was possible to take a walk from any point in the town, cross each bridge once and only once, and return to the starting point. No one had found a way to do it but, on the other hand, no one could explain why it seemed to be impossible. Euler became intrigued by the debate, turning it into a puzzle:

In the town of Königsberg, is it possible to cross each of its seven bridges over the Pregel River, which connect two islands and the mainland, without crossing over any bridge twice?

It is sufficient to say here that it is impossible to make the crossings without doubling back. The bridges of the town form a network with odd vertices in it. This means that the network cannot be traced by one continuous movement without having to double back over paths that have already been traced. Consider an even vertex with two paths going into it, A and B. One can reach the vertex via A and exit via B or vice versa. On the other hand, an odd vertex with three paths,

A, B, and C, involves doubling back. If we reach the vertex via A and exit via B, we can traverse path C only by doubling back via B to the vertex and then exiting through C.

Aware of the capacity of puzzles to model abstract situations cleverly, the nineteenth British mathematician and logician Augustus De Morgan, who wrote important works on the calculus and symbolic logic, produced an ingenious and challenging collection, titled *A Budget of Paradoxes* (De Morgan 1954), in which he illustrated a host of mathematical theories, ideas, and suppositions that were being bandied about at the time. In the same time frame, the *Ladies' Diary*, or *Woman's Almanac*, reflected a new perception of puzzles as part of an ever-broadening leisure culture. Founded in 1704, and initially consisting of recipes, portraits of notable women, and articles on health and education, it soon became the first magazine to include puzzles and riddles for intellectual amusement. The magazine ended publication in 1841. In that same era, Lewis Carroll stepped forward to raise puzzle-making to a veritable art form. He did this by creating puzzles in the form of narratives collected in two books—*Pillow Problems* in 1880 and *A Tangled Tale* in 1886. Carroll is also known for his children's fantasies, *Alice's Adventures in Wonderland* (1866) and *Through the Looking-Glass* (1872). But these, too, are, when looked at more closely, extended narrative puzzles that have amused and challenged children ever since they were first published.

By the late nineteenth century, puzzle-making was becoming a viable profession in its own right. That era saw professional puzzlists come onto the social scene—the Americans Oswald Veblen and Sam Loyd, the Frenchman François Edouard Anatole Lucas, and the Englishman Henry E. Dudeney. Veblen's works became popular because they refashioned classical puzzles into simple paraphrases for mass consumption. Lucas (a brilliant mathematician) is perhaps best known as the inventor of the Towers of Hanoi Puzzle (Lucas 1882):

> *A monastery in Hanoi has three pegs. One holds 64 gold discs in descending order of size—the largest at the bottom, the smallest at the top. The monks have orders from God to move all the discs to the third peg while keeping them in descending order. A larger disc must never sit on a smaller one. All three pegs can be used. When the monks move the last disk, the world will end. Why?*

The reason "the world will end" is that it would take the monks $2^{64}-1$ moves to accomplish the task God set before them. Even at one move per second (and no mistakes), this task will require 582,000,000,000 years. Strikingly, this result recalls Ibn Khallikan's chessboard grain puzzle—suggesting an archetypal principle at work (as will be discussed throughout this book). Sam Loyd produced over 10,000 puzzles in his lifetime, most of which are extremely challenging. Henry Ernest Dudeney was Loyd's British counterpart. In 1893, the two started a correspondence. But this was soon broken off when Dudeney started suspecting Loyd of plagiarizing his ideas.

In the latter part of the nineteenth century, mathematician W. W. Rouse Ball produced the first extensive and in-depth treatment of famous mathematical puzzles. His *Mathematical Recreations and Essays* was first published in 1892, and has undergone at least thirteen editions since. In the subsequent century, Martin Gardner and Raymond Smullyan, made puzzles highly popular through their writings. Between 1956 and 1981, Gardner wrote about the history and meanings of puzzles extensively in a column for *Scientific American*. Smullyan was a logician who composed a host of ingenious chess and logic puzzles designed to strip down logical thinking to its bare essentials. The twentieth century also witnessed a massive proliferation of interest in puzzles among the general populace. The crossword puzzle, invented in 1913, the Rubik's Cube in 1975, and Sudoku and its derivatives around 2005, all became fads tied to recreational culture. Puzzle associations and magazines were established throughout the world. The best known one is probably the Slocum Puzzle Foundation founded by Jerry Slocum in 1993. With Jack Botermans, Slocum put together a compilation of puzzles and games that has became a basic framework for enigmatology (Slocum and Botermans 1994). Puzzle toys and games designed for children also became popular. In the twenty-first century, video games and online venues providing all kinds of puzzle activities, have created a virtual puzzle culture all of its own. As well, research on puzzles, aging and dementia has become a trend in neuroscience and cognitive psychology, reflecting the belief that puzzles stimulate the mind and help regenerate brain cells (Restak and Kim 2010). This belief seems to have substance. The very fact that puzzles have existed across cultures and across time strongly suggests that they are connected to the structure of the brain.

Main puzzle types

To discuss puzzles in any coherent manner, it is useful to have a system of categorization of the main types. This requires a high degree of condensation and amalgamation, given the large number of distinct puzzle genres. In the area of mathematics alone, as the table of contents in Fred Schuh's (1968) classic collection shows, there are at least 267 distinct puzzle types.

A practical strategy is to divide puzzles into general categories—*verbal*, *nonverbal*, *mathematical*, *logical*, and *games*—and then divide them further into subcategories. In the *verbal* category, one would include riddles, anagrams, acrostics, doublets, word searches, crosswords, cryptograms, and the like. The *nonverbal* category could contain visual and geometrical puzzles, along with optical illusions and rebuses (although these could easily be listed in the verbal category as well). Under the *mathematical* rubric, one would put puzzles involving arithmetic, graph puzzles, and the like. *Logical* puzzles would include deduction puzzles, paradoxes, Sudoku (and its derivatives), and so on. The *games* category might include the 14/15 puzzle, the Rubik's Cube, all assembly puzzles, chess, checkers, card games, and the like.

Jerry Slocum's database and collection of puzzles at the Indiana University Library—the most comprehensive in the world—contains over 30,000 puzzles and 4,000 puzzle-related books. David Singmaster (1996, 2004) has provided a chronology of puzzles throughout the ages that could be used as a general typology (see also Wells 1992; van Delft and Botermans 1995; Danesi 2002; Nuessel 2011, 2013). Another way to classify puzzles is according to the psychological strategies employed to solve them—trial-and-error, association, inference, and so on.

One could also classify puzzles according to a hypothetical scale from "0" to "1", with "0" indicating that reasoning is involved and "1" imagination, as will be discussed in the final chapter. It can be said that a puzzle that requires a lot of imagination, such as the Nine-Dot one, falls near the "1-point" end; a puzzle like Sudoku which requires mainly reasoning, can be located instead closer to the "0-point." Finding a point on the scale is problematic, of course, since it involves subjectivity. As Charles Trigg (1978: 21) has aptly observed, determining what puzzles are to be included under a rubric is virtually impossible, because "tastes are highly individualized, so no classification of particular mathematical topics as recreational or not is likely to gain universal acceptance." An amalgam of these classificatory approaches and categories will be used in this book, allowing for significant overlap among them.

Solving puzzles

As far as can be determined, no other animal species displays a comparable need for riddles, puzzles, and intellectual games. They reveal a dialectic structure to consciousness, which aims to probe the nature of things by posing questions about them, and seeking answers wherever and whenever possible. Every puzzle engages us in an internal dialectic with its contents, since it impels us to find an answer. When an answer is not forthcoming, we are left in a frustrating state of mind that can be relieved only when the answer is eventually found.

In all open puzzles (see above), the starting point is to develop a hunch. The American mathematician and philosopher, Charles Sanders Peirce (1931 and subsequent volumes) saw hunches as the spark for solving problems and for grasping new ideas; after that, reasoning takes over to complete the solution. This suggests a "flow model" of puzzles, as already discussed, that moves from the imagination and hunches to reasoning. It can be represented in a diagram as shown below:

Puzzle Statement
Q → (A)

Flow Model
Q → imagination → hunch → Aha insight → reasoning → A

Hunches are initial attempts of the imagination to grasp what something unknown means. As if in a dream, when the insight comes it then suggests a methodical system of reasoning with which to complete the solution. Of course, other processes or phases might be involved. But, by and large, this flow model is sufficient to characterize the dialectic process involved. Hunches often occur through analogies, inferences based on previous experiences, and other contextual factors. Indeed, the expert solver uses these to derive the hunches.

Puzzles are miniature blueprints of how dialectic thinking unfolds. This is arguably why they are universal artifacts of human ingenuity. As Henry Dudeney (1958: 12) aptly put it:

> The curious propensity for propounding puzzles is not peculiar to any race or any period of history. It is simply innate . . . though it is always showing itself in different forms; whether the individual be a Sphinx of Egypt, a Samson of Hebrew lore, an Indian fakir, a Chinese philosopher, a mahatma of Tibet, or a European mathematician makes little difference.

Like other areas of human imagination, puzzles are both "serious" and "artful;" that is, they may be presented as harboring some intrinsic insight or else they may involve trickery and cleverness. Take as an example of the latter a classic puzzle that turned up in an arithmetic textbook written by Christoff Rudolf in Nuremberg in 1561 (Degrazia 1981):

A snail is at the bottom of a 30-foot well. Each day, it crawls up 3 feet and slips back 2 feet. At that rate, when will the snail be able to reach the top of the well?

This puzzle sets a trap, warning us of the dangers of making assumptions. Since the snail crawls up 3 feet, but slips back 2 feet, its net distance gain at the end of every day is, of course, 1 foot up from the day before. To put it another way, the snail's climbing rate is 1 foot up per day. At the end of the first day, therefore, the snail will have gone up 1 foot from the bottom of the well, and will have 29 feet left to go to the top (remembering that the well is 30 feet in depth). If we conclude that the snail will get to the top of the well on the twenty-ninth day, we will have fallen into the puzzle's trap. Consider the start of day twenty-eight. The snail finds itself at 27 feet from the bottom, which it had reached after twenty-seven full days of climbing. This means that the snail has 3 feet to go to the top on that day. It goes up the 3 feet, reaches the top, and goes out, precluding its slippage back down. So, it takes the snail twenty-eight and a half days, because it does not slip back at night on the twenty-eighth day. Rather than an Aha effect, a puzzle of this kind can be said to produce a Gotcha effect, as Martin Gardner (1982) has aptly designated it.

But even though it falls into the artful category of puzzles, like many puzzle genres it, nonetheless, harbors within it what has been called here a puzzle archetype, which is why it surfaces in other puzzle statements. One of these is found in the third section of Fibonacci's *Liber Abaci*. He states it as follows:

A lion trapped in a pit 50 feet deep tries to climb out of it. Each day he climbs up 1/7 of a foot: but each night slips back 1/9 of a foot. How many days will it take the lion to reach the top of the pit?

Both genres—Aha and Gotcha puzzles—reveal the duality of the human mind. As Stanislas Dehaene (1997) has argued cogently, the human brain does not work like a computer; it has evolved to explain the world in its own unique ways. These ways involve both seriousness and trickery.

Puzzles emerged at the same time as myths and the ancient mystery cults, at the dawn of human history. This is not a mere coincidence. Puzzles and mysteries are intrinsically intertwined, generating the same kind of feeling of suspense that calls out for relief through some resolution. The word *catharsis* was used by Aristotle to describe the "emotional relief" that results from watching a tragic drama on stage. Unraveling the solution to a mystery story or to a puzzle seems to produce a similar kind of "mental catharsis," since we typically feel a sense of relief from suspense when we find the answer.

Puzzles thus lend themselves as ideal research materials for investigating the mind. Strangely, there are few studies in psychology or neuroscience that aim to study the relation of puzzles to various aspects of cognition. As far as can be told, the first ever investigation of puzzle-solving from a psychological perspective dates back to an 1897 article by Ernest H. Lindley, titled "A Study of Puzzles with Special Reference to the Psychology of Mental Adaptation." Essentially, Lindley provides an overview of the importance of play in childhood as a stimulant of mental development, prefiguring a whole series of studies on play and games in childhood, of which the ones by Piaget (1969) and Vygotsky (1961) are the best known. Vygotsky proposed developmental stages that go from external (physical and social) actions to internal cognitive constructions and interior speech via the mind's ability to construct images of external reality. His definition of speech as a "microcosm of consciousness" is useful here, since it can be extended to puzzles generally, and especially to riddles. Action, imagination, and abstract thought are the chronologically related stages through which each child passes on the way to mature thinking: that is, the child at first employs the imagination to carry out actions, proceeding to the use of expressive constructs such as single words standing for concepts, and finally using language and other expressive codes that underlie the emergence of abstract thought.

This developmental flow resembles the cognitive flow described above in the solution of puzzles, suggesting that puzzle-solving is a small-scale model of cognitive growth. Loftus and Loftus (1983) extended Vygotsky's model to the study of ludic activities generally—a domain of investigation that has recently included video games (Madigan 2015). Steven Johnson (2005), for example, argues that video games may actually be producing powerful new forms of intelligence, since they provide a channel for the same kind of rigorous mental workout that mathematical theorems do. As a consequence, they improve the problem-solving skills of players, because of a "Sleeper Curve," as Johnson anecdotally calls it. He took the term from Woody Allen's 1973 movie *Sleeper*, in which a granola-eating New Yorker falls asleep, waking up in the future, where junk food and rich foods actually prolong life, rather than shorten it. According to Johnson, video games are turning out to be "cognitively nutritional" after all. Whether or not Johnson's claim is sustainable empirically, it, nonetheless, falls into a line of inquiry that is useful for the present purposes.

This line is based on a critical question: What occurs in the mind as we solve riddles, math puzzles, and the like? Joyce E. Mather and Linus W. Kline were likely the first to tackle this question in 1922, in a paper titled "The Psychology of Solving Puzzle Problems." The authors presented an in-depth discussion of how puzzles are solved by a form of intelligence that is unique among species. Indirectly, this idea had surfaced already in ancient times. The Riddle of the Sphinx was a sort of mythical "intelligence test," revealing the power of the human imagination. To this day, in fact, skill at puzzle-solving is thought to be the privilege of those who, like Oedipus, are endowed with superior intelligence. In a study published in 1982, the psychologists Sternberg and Davidson investigated this very belief, examining the relationship between puzzle-solving and IQ, finding that success at puzzle-solving did not correlate with IQ. No study since theirs has come forward to refute or seriously contradict their study. The two psychologists subsequently continued to pursue this important line of inquiry (Sternberg 1985; Sternberg and Davidson 2003). Simply put, their research suggests that the concept of IQ, and even of intelligence itself, requires revision or at least some elaboration, given that the phenomenon of puzzles raises many questions about our views of intelligence.

Of relevance is the study by Lee, Goodwin, and Johnson-Laird on the psychology of Sudoku (2008). The experiments reported in the study show that there is more to Sudoku than just deductive reasoning. A major strategic shift in cognition is necessary to acquire tactics for solving more difficult (complex) puzzles, since solvers have to keep track of possible digits in a cell. The study suggests that reasoning is crucial in solving puzzles but that there are other aspects that involve memory and inferential thinking. Puzzles, in effect, provide a unique opportunity to examine the construct of intelligence in a unified manner.

Puzzle archetypes

Let us look more closely at Alcuin's River-Crossing Puzzle. Actually, there are four such puzzles in the *Propositiones*, numbered 17, 18, 19, and 20. They are reproduced here for the sake of convenience (see Hadley and Singmaster 1992; Burkholder 1993; Singmaster 1998):

> Number 17: *Propositio de tribus fratribus singulas habentibus sorores*
>
> *There were three men, each having an unmarried sister, who needed to cross a river. Each man was desirous of his friend's sister. Coming to the river, they found only a small boat in which only two persons could cross at a time. How did they cross the river, so that none of the sisters were defiled by the men?*
>
> Number 18: *Propositio de homine et capra et lupo*

A certain man needed to take a wolf, a she-goat and a load of cabbage across a river. However, he could only find a boat which would carry two of these [at a time]. Thus, what rule did he employ so as to get all of them across unharmed?

Number 19: *Propositio de viro et muliere ponderantibus [plaustri pondus onusti]*
A man and his wife, each the weight of a loaded cart, who had two children each the weight of a small cart, needed to cross a river. However, the boat they came across could only carry the weight of one cart. Devise [a way] of crossing in order that the boat should not sink.

Number 20: *Propositio de hirtitiis*
A masculine and feminine [. . .] who had two children weighing a pound wished to cross a river.

Puzzle 20 is obviously incomplete and thus will not be discussed here. Puzzle 17 is fascinating on several counts. Culturally, the wording of this puzzle provides a snapshot of gender relations in the medieval era—today, it would sound antiquated or anomalous (or perhaps even sexist) in its portrayal of such relations. Mathematician Niccolò Tartaglia became especially intrigued by this puzzle, inventing one of his own in which three brides and their jealous husbands had to go across the river—another culturally revealing wording of the same puzzle archetype (see O'Beirne 1965: 4–5). In Tartaglia's version, the only stipulation is that no wife can be left on either side or on the boat without the presence of her own husband. Nine back-and-forth trips are required to solve this version of the puzzle. Interestingly, it is impossible to arrive at a solution under the conditions stipulated by the puzzle for four couples, unless there is an island in the middle of the river, which can be used as a temporary landing place. In other cases, the island affects neither the total number of trips nor the feasibility of a successful transfer (Loyd 1959: 131–2).

Puzzle 18 is the main one—the others being essentially derivatives and more complex variations of the basic situation it enfolds (Pressman and Singmaster 1989). The solution hinges on making the first trip over successfully, as already discussed above, which means starting with the goat. Once this critical decision is made, the rest of the puzzle is solved easily, as we saw. That decision derives from examining the possibilities hypothetically in the mind. As such, it displays how the flow theory of puzzles unfolds in a nutshell. Moreover, it harbors within it the ideational nucleus of what subsequently came to be known as critical path theory—the analysis of complex decisions with reference to their outcomes. It constitutes an example of what has been called a puzzle archetype here, which is passed on, eventually sparking a set of related ideas and principles that can be derived from it. The puzzle is also the likely archetype behind the theory of combinatorics, or the study of the structure of arrangements under certain circumstances.

Reinforcing the claim that Alcuin's puzzle is an archetype is the fact that similar river-crossing puzzles have appeared in a variety of cultures across time (Ascher 1990). As Olivastro (1993: 139) comments:

> Simpler versions can be found in several African folktales where they are taken not to be difficult problems but only pleasant stories. In the Swahili tradition, a visitor from another region visits a sultan but refuses to pay tribute. He is confronted with a challenge: He must carry a leopard, a goat and some tree leaves to the sultan's son who lives across a river, and he must use a boat that will hold the visitor and two other items. The problem, of course, is that no two items can be left on the shore together. (This is different from the version mentioned by Alcuin, which gave the option of leaving at least the wolf and cabbage on the shore together.) The visitor, after mulling over the problem, decides to carry first the leaves and goat, return with the goat, and then carry the goat and leopard together to the son.

Olivastro goes on to suggest that Alcuin may have derived his own river-crossing puzzles from other sources. Whatever the truth, the archetype imprinted in the puzzles has surfaced elsewhere where contact between Alcuin and other puzzle-makers would be unlikely, given the different languages and societies in which the versions were constructed. A nineteenth-century version involves three missionaries and three cannibals. The latter must never be allowed to outnumber missionaries on either bank. Again, it takes nine trips to get everyone across safely:

1. A missionary and a cannibal cross over.
2. The missionary returns.
3. The missionary gets off and two cannibals cross.
4. One cannibal gets off and the other one returns.
5. The cannibal gets off and two missionaries cross.
6. One missionary gets off and the other missionary returns.
7. The two missionaries cross.
8. The two missionaries get off and one cannibal returns.
9. The remaining cannibals cross.

As these examples show, the culturally idiosyncratic details of the archetype might vary, but its underlying structure is always the same. As Martha Ascher

(1990: 26) aptly observes, the different cultural versions of the puzzle "are expressions of their cultures and so variations will be seen in the characters, the settings, and the way in which the logical problem is framed." An interesting variation of Russian origins is paraphrased below:

> *Three soldiers have to cross a river without a bridge. Two boys with a boat agree to help the soldiers, but the boat is so small it can support only one soldier or two boys. A soldier and a boy cannot be in the boat at the same time for fear of sinking it. Given that none of the soldiers can swim, it would seem that in these circumstances just one soldier could cross the river. Yet, all three soldiers eventually end up on the other bank and return the boat to the boys. How do they do it?*

The crossings are organized as follows:

1. Both boys go to the opposite bank and one gets off.
2. The other boy brings the boat back to the soldiers and gets off.
3. A soldier crosses the river and gets off.
4. The boy there returns with the boat.
5. When he gets to the original side, both boys get on and cross the river.
6. One boy gets off and the other returns with the boat.
7. The boy gets off and a second soldier crosses the river.
8. He gets off and the second boy returns with the boat.
9. Both boys get on and cross the river.
10. One boy gets off and the other returns with the boat.
11. He gets off and the third soldier crosses the river.
12. The soldier gets off and the second boy returns with the boat.
13. The two boys get on and cross over.

Recall Euler's Königsberg's Bridges Puzzle above, which also involves traversing paths (bridges in this case), in order to determine if it is possible to do so minimally without doubling back on any of the paths. The archetype in this case reveals the same abstract structure of Alcuin's puzzles—traveling over paths according to specific conditions, constraints, or rules. It is somewhat amazing to think that the puzzle also led to mathematical discoveries. Euler had formulated the puzzle in a famous 1736 paper that he presented to the Academy in St. Petersburg, Russia, and which he published in 1741 (Richeson 2008). It describes an actual layout of the bridges. To reiterate the statement here for convenience, on the river there are two islands connected with the mainland and with each other by seven bridges. Is it possible to cross each bridge once and only once, and return to the starting point? Euler went on to prove that it is impossible to trace a path over the bridges without crossing at least one of them twice. He used a graph version in which the landforms could be seen to constitute a network of vertices, and the bridges as paths or edges. As we discussed briefly above, a network can have any number of even paths in it, because all the paths that converge at an even vertex are "used up" without having to double back on any one of them. For example, at a vertex with just two paths, one path is used to get to the vertex and another one to leave it. Both paths are thus used up without having to double back over either one of them. In a four-path network, when we get to a vertex, we can exit via a second path. Then, a third path brings us back to the vertex, and a fourth one gets us out. All paths are once again used up. In an odd vertex network, on the other hand, there will always be one path that is not used up. For example, at a vertex with three paths, one path is used to get to the vertex and another one to leave it (as mentioned briefly above). But the third path can only be used to go back to the vertex. To get out, we must double back over one of the three paths. The same reasoning applies to any odd vertex network. Therefore, a network can have, at most, two odd vertices in it. And these must be the starting and ending vertices. The relevant point here is that Euler's brilliant puzzle made it possible to look at the relationships among elemental geometric systems in an abstract fashion in order to determine their structure and their implications. Richeson (2008: 107) puts it as follows:

> The solution to the Königsberg bridge problem illustrates a general mathematical phenomenon. When examining a problem, we may be overwhelmed by extraneous information. A good problem-solving technique strips away irrelevant information and focuses on the essence of the situation. In this case details such as the exact positions of the bridges and land masses, the width of the river, and the shape of the island were extraneous. Euler turned the problem into one that is simple to state in graph theory terms. Such is the sign of genius.

The implications of Euler's puzzle laid the foundations of graph theory, topology, and the mathematical study of impossibility. Graph theory has had a great impact on mathematical methodology itself, bringing together areas that were previously thought to be separate. In other words, the archetype that the puzzle harbors has many other inherent meanings and these have led to concrete discoveries. Puzzle archetypes are not limited to mathematical puzzles. The image of the phases of life as related to phases of the day, in the Riddle of the Sphinx, occurs across many riddle traditions. In Philippine mythology, the sphinx is depicted as part human and part eagle and it, too, asks riddles to wanderers who trespass in the Bicol region. Anyone who fails to answer its riddle is carried off to the Mayon Volcano and offered to the volcano god to appease his anger.

As is well known, the term *archetype* was proposed by Carl Jung (1983) to describe an unconscious figure of mind that finds expression in rituals, symbols, forms, and words. Archetypes are deeply rooted in the psyche deriving from the experience of everyday life events. They are understood in the same way across time and geographic space, although their specific manifestations vary, because they are part of the brain's make-up. Archetype theory explains the recurrence of such ideas as river-crossing and doubling-back situations. As Michael Schneider (1994) has cogently argued, the patterns that recur in the universe are processed archetypally by the mind, including the presence of hexagonal structure in beehives, the manifestation of the spiral form in many natural phenomena, and so on. A puzzle archetype can thus be defined as an abstract image that shows up in different guises across temporal and cultural spaces, but shows the same underlying structure.

Puzzles and human intelligence

If an archetype theory of puzzles is plausible or sustainable, then the question of the relation of puzzles to the origins of consciousness and human intelligence inevitably ensues. As discussed above, riddles and other puzzle artifacts emerge at the same time as the first human cultures do, which, as Jung argued, were built on the basis of common archetypes embedded in our collective unconscious. It is thus little wonder that Sigmund Freud, the founder of psychoanalysis, turned to the myths to extract from them insights into the human mind. He came to the notion of the *Oedipus Complex*, for instance, on the basis of the Oedipus story (Freud 1901), defining it as a feeling of hostility toward the parent of the same sex and an attraction to the parent of the opposite sex, eventually leading to neurotic behavior. The ancient myths and riddles are *ipso facto* theories of mind.

In this regard, it is useful to consider the notion of "imaginative universals" of the seventeenth–eighteenth century Italian philosopher Giambattista Vico (in Bergin and Fisch 1984). An imaginative universal is the precedent to archetype formation—it consists in notions that we all sense as intuitively meaningful. The Sphinx's Riddle is based on one such universal—imagining the phases of day as a metaphor for the phases of life. The psychologist Julian Jaynes (1976) used a similar argument to reconstruct the original form of consciousness—an argument based on the interpretation of myths and symbols. If right, then there is little doubt that the ancient riddles were products of an impetus to express the archetypal forms of the imagination in a concrete way. And this is arguably why they are found at the origins of cultures across the world and across languages.

Puzzle archetypes take on a particular structure according to culture and time, as we saw with the river-crossing puzzles above. Puzzles put the powerful abilities of human intelligence on display. This is not a measurable process, as implied in IQ tests; it is a mode of consciousness guided by the hunches that spring from the imagination. Mathematician Norbert Wiener (1948, 1950) was among the first to draw analogies between machines, animal minds, and human minds, viewing the different forms of intelligence as analogous. Because of the increasing sophistication of computers and the efforts to make them behave in humanlike ways, Wiener's perspective today is closely allied with artificial intelligence and robotics. As useful as this perspective is, it does not take into account the difference between conscious and unconscious thought. Animals exchange signals, interpreting them in terms of affective and survival mechanisms. Machines exchange programmed information acting on it computationally. Only humans exchange information that requires interpretation that taps into unconscious structures of cognition—the same structures that underlie puzzle archetypes.

The Riddle of the Sphinx is one the first examples in human history of how a puzzle archetype betrays the nature of human intelligence. Its origin in myth resonates to this day in the stories composed for children. The heroes in such stories typically face challenges designed to test not only their physical mettle, but also their ability to solve riddles and other puzzles. As such narrative traditions suggest, we perceive riddles as "miniature revelations" of truth. What are philosophy and science, after all, if not attempts to answer the riddles that life poses? Mathematical inquiry, too, seems to be guided by an inborn need to model perplexing ideas in the form of archetypal puzzles. This is perhaps why some of the greatest questions of mathematical history were framed originally as puzzles. Solving them required a large dose of insight thinking. In many cases, the insight took centuries and even millennia to come. But eventually it came, leading subsequently to intellectual progress. It would seem that in order to grasp the meaning of the "Thebes" of human understanding, we must first metaphorically solve the Riddle of the Sphinx, or the riddle of human intelligence.

2

Riddles

It is of first-class importance that our answer to the Riddle of the Sphinx should be in step with how we conduct our civilization, and this should in turn be in step with the actual workings of living systems.

Gregory Bateson (1904–1980)

The ancients took riddles seriously, as many early legends attest. The mythical figures of yore earned their heroic status not only through their physical prowess, but also (if not more so) through their ability to solve challenging riddles. The hero Oedipus did so, however, at his own unwitting risk, as we saw. By so doing, he gradually became consciously aware of the calamity that awaited him, fulfilling the ominous prophecy of the oracle who had proclaimed his tragic destiny to his father at birth. No doubt, the unrecorded oracle's prophecy was also couched in the form of a riddle.

Stories of this type abound in antiquity. The one of the Hebrew hero Samson is another one. At his wedding feast, Samson wanted to impress the relatives of his wife-to-be by posing the following riddle to his Philistine guests (Judges 14: 14):

Out of the eater came forth meat and out of the strong came forth sweetness.

He gave the Philistines seven days to come up with the answer, convinced that they were intellectually incapable of doing so. He promised them a reward of thirty items of clothing should they come up with the right answer. The clever Samson had devised his riddle as a description of something he had once experienced. Apparently, he had killed a young lion and subsequently saw bees making honey in its carcass. The villainous Philistines, however, took advantage of the time given to them to threaten Samson's wife, coercing her to get the answer from her husband and pass it on to them: the "eater" = "swarm of bees;" the "lion" = "the strong;" and "made honey" = "came forth sweetness." When the Philistines announced the correct response, the mighty Biblical hero became enraged, seeking to avenge himself against all Philistines. The ensuing conflict led eventually to Samson's own destruction. As in the Oedipus story, the calamity was associated with a riddle.

The riddles of the Sphinx and of Samson are examples of how they tell a fascinating side-story about human destiny. The inability to solve a riddle was believed to bring about dire consequences to heroic figures. To the list of casualties, one can add the ancient Greek poet Homer, whose death was said to have been precipitated by the distress he felt at his failure to solve the following riddle posed to him by a group of fishermen:

What we caught, we threw away. What we could not catch, we kept (answer: fleas).

While the repercussions of not solving riddles today are not as catastrophic, it is nonetheless true that failure makes us feel less than "intellectually heroic." Children in particular are instinctively drawn to riddles, clearly enjoying both the challenge they pose and the mischievous language with which they have been constructed. The fact that no one has ever explained to a child what a riddle is constitutes persuasive evidence that the brain is a ludic organ. As Hovanec (1978: 12) aptly remarks:

> Riddles would have to be considered the oldest form of puzzles. They are to be found throughout the mythologies of many cultures and were important in ancient religion and philosophy as one of the vehicles through which the sages expressed their wisdom. Primitive peoples were obsessed with imagery and looked upon the ability to pose and solve riddles as indicative of mental agility.

Not all early riddles were associated with mythic stories and with prophecies. As briefly mentioned, Solomon and Hiram organized riddle competitions, the Romans made riddles a recreational activity during the Saturnalia, and by the fourth century CE riddles had, in fact, become so popular for their recreational value that memory of their mythic origins started to fade. By the tenth century, after Arabic scholars used riddles for pedagogical reasons—namely, to train students in the art of detecting linguistic ambiguities, coinciding with the establishment of the first law schools of Europe—riddles assumed many new ludic social functions.

This chapter explores these functions and the relation of riddles to human intelligence. As Johan Huizinga persuasively argued in his classic book, *Homo Ludens* (1938), the structure of cultures throughout the world is essentially ludic, even though this may not be apparent because over time rites, symbols, and rituals are no longer perceived as originating in a form of play.

Riddles as oral tradition

Riddles emerge as part of oral traditions, likely recited by elders or oracles in early cultural settings, and later by designated orators. Many of the riddles that became part of a culture's folklore traditions were eventually written down, with the advent and spread of writing, but even so they were orated or told at gatherings—given low literacy rates. The difference between a riddle as told by an oracle and one by an orator is found in its social function; the former was intended as a cautionary, divinatory, or wise tale; the latter to entertain people in clever and ingenious ways.

The Riddle of the Sphinx is one of the earliest written riddles, going back to around 2500 BCE by most estimates. As discussed, it reflects the divinatory function of early riddles. A similar kind of ancient riddle was found inscribed on a tablet, dating back to Babylonian times around 2000 BCE (Seyeb-Gohrab 2010: 14). Some of the oldest surviving riddles are, actually, in the Sanskrit *Rigveda* and various Vedic texts (Taylor 1948; Salomon 1996). The great Hindu epic, the *Mahabharata* (*c.* 400 CE), includes riddles and describes riddle contests, not unlike those of ancient Greece and Rome (Kaivola-Bregenhøj 2001). According to Archer Taylor (1948: 28):

> the oldest recorded riddles are Babylonian school texts which show no literary polish. The answers to the riddles are not preserved; they include my knees hasten, my feet do not rest, a shepherd without pity drives me to pasture (a river? a rowboat?); you went and took the enemy's property; the enemy came and took your property (a weaving shuttle?); who becomes pregnant without conceiving, who becomes fat without eating? (a raincloud?). It is clear that we have here riddles from oral tradition that a teacher has put into a schoolbook.

In fact, the archeological record indicates that riddles had always had both divinatory and recreational-educational functions from the start of civilization.

The early medieval period saw a flourishing of riddle collections across every part of the world. The fact that they were compiled suggests that they had become staples of folk culture, probably recited by someone after, say, dinner, in a village square setting. The orator had high social status, having the ability to read them out loud effectively and dramatically. One of the most famous collections comes from the fourth–fifth century CE, consisting of 100 riddles in the form of brief Latin epigrams. It was written by a certain Symphosius, about whom virtually nothing is known. It is called the *Aenigmata Symphosii* (the enigmas of Symphosius). The following is a typical example of his riddle art (Hovanec 1978: 14–15):

The Mule
Unlike my mother, in semblance different
from my father, of mingled race, a breed
unfit for progeny, of others am I born, and
none is born of me.

The fact that the answer is given at the start suggests that the riddles were spoken by someone who could read and, knowing the answer, would be better able to dramatize the riddle. Many of these riddles were reminiscent of the ancient fables of Aesop, dealing with animals as human metaphors for character. The preface to a subsequent edition of the collection states that "within Symphosius' milieu there is still a conception of riddles as oral and agonistic." Erin Sebo (2009) argues that the riddles are framed with a high literary quality, establishing Symphosius' collection as the first one transforming riddling into a literary art. A little later, the English scholar and poet Aldhelm (640–709 CE), produced a collection of 100 riddles, titled *Aenigmata,* which influenced subsequent riddle styles. Aldhelm's riddles put on display the power of metaphorical language to imbue everyday things (animals, plants, household items, and so on) with social and metaphysical meanings. The following riddle, for example, is a vivid metaphorical portrait of the nature of dogs as seen through human eyes (Hovanec 1978: 15):

The Dog
Long since, the holy power that made all things
So made me that my master's dangerous foes
I scatter. Bearing weapons in my jaws,
I soon decide fierce combats; yet I flee
Before the lashings of a little child.

Less well known are the riddles of Alcuin (Chapter 1). Unlike Symphosius and Aldhelm, Alcuin did not give his riddles a title, likely because they were not meant to be orated. Rather, he included them in his letter correspondences (Lapidge and Rosier 1985). The following one, which he sent to the Archbishop of Mainz, known by the nickname of Damoeta, is a typical example (Hovanec 1978: 15):

A beast has sudden come to this my house,
A beast of wonder, who two heads has got,
And yet the beast has only one jaw-bone.
Twice three times ten of horrid teeth it has.

Its food grows on this body of mine,
Not flesh, nor fruit. It eats not with its teeth,
Drinks not. Its open mouth shows no decay.
Tell me, Damoeta dear, what beast is this?
(answer: a comb)

The solution to this riddle hinges on decoding the figurative meaning of *beast* as a device rather than an animal. Referring to a comb as a beast and describing it as if it were one allows Alcuin to convey his negative feelings about this object cleverly and to project insights into the nature of human-made objects as having anthropomorphic power.

Riddles may, in fact, be defined as extended metaphors, designed to provide figurative insights into the nature of things through the power of language to connect things by association. In other words, the path to A in the Q → (A) structure is via metaphorical reasoning. This is perhaps why riddles are so effective, revealing how the human mind is inclined to understand something in terms of something else. Each riddle is akin to a condensed poem. The anonymous *Exeter Book* (*c.* 960), contains nearly 100 riddles that were, in fact, composed as miniature poems about everyday mundane things such as storms, ships, beer, books, and falcons. They are composed in the style of medieval Anglo-Saxon poetry. Indirectly, the book constitutes a portrait of medieval English life with its inherent values and anxieties, such as the meaning of suffering and the passage of time (Chambers, Förster, and Flower 1933). The same type of poetic style for characterizing everyday life is evidenced across riddle traditions. For example, the collection by the Persian Amir Khusro (1253–1325), written in an Indic language called Hindawi, contains 286 riddles divided into six themes that covered concepts similar to those in the *Exeter Book*, and which can likewise be solved by making figurative links among the words and the concepts (Vatuk 1969).

In the tenth century, a number of Arabic scholars used riddles specifically to alert people to the dangers that this kind of language, with all of its ambiguities, *double entendres*, and semantic possibilities, poses for clear communication (Scott 1965). Al-Hariri of Basra (*c.* 1050–1120), for example, composed riddles for his *Assemblies*, in which he also discussed problems of grammar (Shah 1980). The riddles were intended to illustrate to students of jurisprudence how metaphorical style can introduce ambiguities and misunderstandings into legal discourse. In the fourteenth century, riddling took another ludic turn. A Benedictine monk named Claret became quite famous among the monks and aristocrats of the era, the main groups who had literacy, by exploiting double entendres, creating quasi-obscene riddles intended presumably as vehicles of prurient entertainment (Ziolkowski 1998). Here is a typical example of Claret's crafty riddling style (Hovanec 1978: 22):

A vessel have I
That is round like a pear,
Moist in the middle,
Surrounded with hair;
And often it happens
That water flows there.
(answer: an eye)

The image that the impish monk obviously wanted to elicit in his reader was a transparent one, although the actual answer of an "eye" also fits the riddle's metaphorical description. This kind of riddling was, clearly, a way to get around the moralistic strictures imposed by the Church on language. By the late Renaissance, riddles were being tailored more and more to produce humorous or whimsical effects, becoming connected with a secular leisure culture and more and more removed from their ancient use as portentous warnings. One well-known English collection of the era, *The Merry Book of Riddles*, was published in 1575. Below is an example from that work (Schiltz 2015):

He went to the wood and caught it,
He sate him downe and sought it;
Because he could not finde it,
Home with him he brought it.
(answer: a thorn caught on a foot or optionally lice)

The eighteenth century saw riddles morph into a popular literary genre, included as regular features in newspapers and periodicals (Taylor 1951). Writers, poets, and philosophers constructed riddles to probe the practical nature of things, recalling Symphosius and Aldhelm. Benjamin Franklin composed riddles under the pen name of Richard Saunders for inclusion in his *Poor Richard's Almanack*, first published in 1732. The section of "mathematical exercises" was a popular feature in the almanac's unexpected success. These were math puzzles that Franklin composed in the form of riddles, adding to their ingenuity and appeal. In France, riddles became a side interest for a number of literary figures, including Voltaire, who would regularly compose ingenious riddles, such as the following one (Hovanec 1978: 28):

What of all things in the world is the longest, the shortest, the swiftest, the slowest, the most divisible and most extended, most regretted, most neglected, without which nothing can be done, and with which many do nothing, which destroys all that is little and ennobles all that is great? (answer: time)

As Voltaire knew, describing time as *long*, *swift*, *slow*, *divisible*, *extended*, and so on is so common in everyday discourse that we are largely unaware that these expressions are part of a conceptual formula that envisions *time* as a physical entity. A riddle such as this one is a kind of treatise on the nature of language.

The ever-increasing popularity of riddles in the nineteenth century led to the invention of social games based on them. One of these was the *mime charade*, which is still played today. Members of separate teams act out the various syllables of a word or phrase in pantomime. If the answer to the charade is, for example, *football*, the words *foot* and *ball* are the ones normally pantomimed. The charade became a social fad among the middle classes, as evidenced by its inclusion in nineteenth-century novels such as William Thackeray's *Vanity Fair* (1848) and Charlotte Brontë's *Jane Eyre* (1847). One of the most famous ones was composed by Jane Austen (Doody 2015: 8):

When my first is a task to a young girl of spirit,
And my second confines her to finish the piece,
How hard is her fate! but how great is her merit
If by taking my whole she effects her release!
(answer: hem-lock)

The literary charade made its way to America through the *Penny Post*, a magazine founded in Philadelphia in 1769. An American collection, based on familiar household objects titled *The Little Puzzling Cap*, was published in 1787. By the nineteenth century, charades and riddles generally had become firmly embedded in American childhood and literary recreational culture. By the twentieth century (and certainly in the twenty-first), riddles had lost some of their recreational appeal, giving way in popularity to other puzzles such as crosswords, Sudoku, and video games. Riddles are now used primarily in popular culture and children's literature. For example, they are part of the villainous discourse of the Riddler (in the Batman comics and movies) and a plot element in J. R. R. Tolkein's *The Hobbit* (1937). The fading of riddling as a literary art is a sign of the times, indicating that the art may have lost its previous social functions. In some ways, riddles are now artifacts in a literary museum, where other literary genres and styles similarly reside. While they may no longer resonate with contemporary society, in the ways that they did previously, they are nevertheless testaments of how human intelligence may have originated and evolved over a relatively short period of historical time.

Riddles have historically been one of the channels through which we have come to a poetic understanding of reality (Burrows 2014). As Elli Köngäs Maranda (1976: 131) has so aptly put it, discussing Malaitian riddles: "riddles make a point of playing with conceptual boundaries and crossing them for the intellectual pleasure of showing that things are not quite as stable as they seem." Various Hebrew stories tell of how the Queen of Sheba tested Solomon's wisdom and intellect with riddles (I Kings 10.1–13, II Chronicles 9.1–12). The Indian *Mahabharata* riddles are posed by a nature spirit to several heroic characters who then act accordingly. Analogously, in the anonymous *100 Riddles of the Fairy Bellaria*, published in 1892, a queen named Bellaria and her riddle-solving skills are pitted against a cruel invading king named Ruggero, who gives her 100 riddles to solve. Ruggero threatens to destroy her empire should she fail to solve them successfully. The same use of riddles as intellectual tools or weapons appears in contemporary movies as disparate as *Harry Potter and the Goblet of Fire* (2005), the *Saw* series of films, and *Die Hard with a Vengeance* (1995).

In some contemporary fiction the mythical sense of riddling shapes the entire narrative. A salient example is Patricia A. McKillip's *The Riddle Master* trilogy (1999) in which the ancient art of riddle-making is taught and studied at the College of Caithard, based on books recovered from the ruins of the ancient School of Wizards. The riddles are constructed with three components: the Question, the Answer, and the Stricture—a characterization that reflects the dialectical structure of riddles discussed in this book. Evoking many of the ancient themes of riddling, the series brings the importance of riddles into contemporary focus, by emphasizing that the wisdom of the ancients must not be lost, if we are to avoid a catastrophic loss of civilization.

Imparting wisdom to children has always been a primary function of riddle traditions. Lewis Carroll certainly understood this, incorporating both actual riddles into his stories and designing the stories themselves as over-arching riddles. *Alice in Wonderland* is an imaginative realm (Wonderland) in which riddles constitute the main language. In that novel, there is a particularly interesting riddle that the Mad Hatter poses to Alice: "Why is a raven like a writing desk?" When Alice gives up on finding an answer and the Mad Hatter admits that he does not know the answer either, we are left in an intellectual quandary, seeking an answer (as if a vital clue to our existence rested on it). This was probably Carroll's intent, conveying to readers how riddles make us reflect on things. Unable to leave Carroll's riddle unanswered, many have subsequently come forth with solutions. One of the most famous is by Aldous Huxley in a

1928 article he wrote for *Vanity Fair*: "Because there is a B in both and an N in neither." Not only do we not envision this as an answer, but it, too, seems to constitute a riddle.

Solving riddles

Consider two of Gollum's riddles in Tolkein's *The Hobbit* (1937):

Riddle 1:
What has roots as nobody sees,
Is taller than trees,
Up, up it goes,
And yet never grows?

Riddle 2:
Voiceless it cries,
Wingless flutters,
Toothless bites,
Mouthless mutters.

The answer to the first riddle is a *mountain* and to the second one the *wind*. Both involve a figurative play on the meanings of words. The first one describes features of mountains that may escape attention, depicting them in a figurative way. When the images are united, we can actually "see" the mountain come into mental view. The riddle reveals, in its miniature way, that understanding occurs by connecting images metaphorically. The second riddle brings us directly into the realm of figurative cognition, grafting different sensory experiences associated with the wind, uniting them into a metaphorical snapshot of what the sensation of the wind evokes in us.

By and large, all riddles show a similar structure and intent—to show hidden connections among things. This play on associative language is even more evident in riddles known as *conundrums* and *enigmas*. The former exploit the similar sounds of word pairs such as "red" and "read," and the different meanings of words or expressions such as "all over" (Hovanec 1978: 28):

What is black and white and red all over? (answer: a newspaper)

Enigmas (from the Greek "to darken and hide") are rhyming riddles that contain one or more veiled references to the answer, as in the following famous enigma composed by the British statesman George Canning (1770–1827):

A word there is of plural number,
Foe to ease and tranquil slumber;
Any other word you take
And add an "s" will plural make,
But if you add an "s" to this,
So strange the metamorphosis;
Plural is plural now no more,
And sweet what bitter was before.
(answer: cares—caress)

One of the most interesting analyses of riddles comes from anthropologist Alan Dundes in 1963, who sees them exploiting oppositional structure (good-versus-evil, light-versus-darkness, and so on), thus connecting them directly with myths, as Claude Lévi-Strauss so cogently argued in 1958. Crucial to Dundes' conception is, therefore, the Saussurean (1916) notion of *value* (*valeur*). Rather than carrying intrinsic meaning, Saussure argued that signs had value only in differential relation to other signs. To determine the value of a riddle, therefore, it must be opposed to something else. So, the Riddle of the Sphinx has *valeur* because it opposes two conceptual domains—the phases of the day against the phases of life. From this opposition, the meaning of the riddle crystallizes spontaneously.

In the *Rhetorica* (1952), Aristotle provides the first definition of riddles as metaphorical speech. In Book III, he states that: "good riddles do, in general, provide us with satisfactory metaphors: for metaphors imply riddles, and therefore a good riddle can furnish a good metaphor." Metaphor is at the core of the riddle-maker's repertoire of word play. Let's take a hypothetical example. The word *laugh* is found in idioms such as *laugh in someone's face* ("to show contempt for someone") and *laugh out of the other side of one's mouth* ("to feel embarrassed after wrongly feeling satisfaction about something"). A riddle that connects these two semantic domains could be the following one:

It can be done to someone's face and can come out of the other side of one's mouth, but it is neither spit or kissing. What is it?

As David Wells (1995: 169) points out, the appeal of riddles such as this one lies in the fact that their language invariably denies words their literal meanings: "To the riddle *What is the difference between a hill and a pill?* the response *One is smaller than the other*, is not acceptable; the correct answer, recognizable by its twist, is *One is harder to get up, the other is hard to get down*." The key to solving riddles lies, as Wells aptly observes, in detecting the metaphorical value that arises when particular words are combined. In isolation, a word can, and does, have literal meaning. But when combined in riddles, literal meaning is denied, via artfully contrived language that creates "a web of antithetical relationships," as J. C. Heesterman (1997: 67) puts it.

The language of riddles is akin to the language of jokes. Like the latter, riddles can be composed on purpose to make fun of something or, simply, to provide a type of comic relief. Take, for example, a classic children's riddle, *Why did the chicken cross the road?* The number of answers to this question is endless:

1. To get to the other side.
2. Because it was taken across by a farmer.
3. Because a fox was chasing it.

All three answers state the obvious, but might escape our attention at first because we have probably chosen to search for a nonobvious answer. The answers tend to provoke moderate laughter, similar to the kind that the "punch line" of jokes might elicit. This is because, as Wells (1988: 7) puts it, riddles and jokes are both dependent "on a skeleton of double meanings, surprise, and the familiar in strange dress." Rather than an Aha Effect, this kind of riddling may thus be said to produce a Ha-Ha Effect (Paulos 1980, 1985).

In 1922, Wittgenstein became fascinated by how language presented information about the world. He saw sentences as propositions about simple world facts. He developed a picture theory of language by which propositions were purported to represent features of the world in the same way that pictures did. But Wittgenstein's theory does not hold up with riddles. Wittgenstein actually had serious misgivings about his own theory. In his posthumously published *Philosophical Investigations* (1953), he was perplexed by the fact that language could do much more than just construct propositions about the world. So, he introduced the idea of "language games," by which he claimed that there existed a variety of linguistic games, including riddles, that went beyond a simple mirroring of the world. Wittgenstein had in effect come to the same realization of the ancients— riddles make us both reflect on the world and often laugh at it through our metaphorical portrayals of that world. As the following Chinese riddle shows, riddles do not only portray things; they shed light on them (Plaks 1996):

千 里 会 千 金

The separate characters mean literally "thousand kilometer meet thousand gold" (Tan, Wei, Dong, Liv, and Zhou 2016). A Chinese proverb states that: "A good horse can run thousands of kilometers per day." So, 千 里 ("thousand kilometer") is resolved as 马 ("horse"). In Chinese tradition, the daughter in a family holds great value. So, it is now possible to resolve 千金 ("thousand gold") as 女 ("daughter"). By blending the radical 马 ("horse") with the radical 女 ("daughter"), the character 妈 ("mother") emerges. This is the answer. This riddle shows how intrinsically intertwined riddles are with cultural *valeur*. Indeed, the connection between riddles and culture is unmistakable.

Riddles and metaphor

The discovery of metaphor by Aristotle came via a contemplation of riddles, as briefly discussed. Aristotle saw metaphor as a product of proportional reasoning—the period of childhood is to life as the morning is to the day; the period of adulthood is to life as the afternoon is to the day; hence, old age is to life as the evening is to the day. As Umberto Eco (1984: 88) aptly pointed out, despite "the thousands and thousands of pages written about metaphor" since Aristotle formulated his theory, no single explanation has ever really eclipsed it. Aristotle actually dismissed metaphor, affirming that, as knowledge-productive as it was, the most common function of metaphor in human life was as a means to spruce up more basic literal ways of speaking and thinking (Aristotle 1952: 34). The Roman Quintilian (*c*. 35–100 CE) subsequently claimed that metaphor was a simple substitutive option to literal speech. Thus, in an expression such as *Julius Caesar is a lion*, Quintilian claimed that we simply substitute the term *lion* for its literal counterpart, a *courageous man*, so as to make it more memorable or effective.

Such views are based on the belief that literal meaning is the default form of language. Already in the medieval period, no less a philosopher than St. Thomas Aquinas went contrary to this literalist grain by claiming, in his *Summa Theologica* (1266–73), that the writers of Holy Scripture presented "spiritual truths" under the "likeness of material things" because that was the only way in which humans could grasp such truths, thus implying that metaphor was a feature of mind, not just a feature of rhetorical flourish (quoted in Davis and Hersh 1986: 250):

> It is befitting Holy Scripture to put forward divine and spiritual truths by means of comparisons with material things. For God provides for everything according to the capacity of its nature. Now it is natural to man to attain to intellectual truths through sensible things, because all our knowledge originates from sense. Hence in Holy Scripture spiritual truths are fittingly taught under the likeness of material things.

But despite St. Thomas' discerning observation, philosophers continued largely to ignore, and even condemn, metaphor. The source of latter view is, probably, John Locke's characterization of metaphor as a "fault" in his *Essay Concerning Humane Understanding* (Locke 1690: 34):

> If we would speak of things as they are, we must allow that all the art of rhetoric, besides order and clearness, all the artificial and figurative application of words eloquence hath invented, are for nothing else but to insinuate wrong ideas, move the passions, and thereby mislead the judgment; and so indeed are perfect cheats: and therefore, however laudable or allowable oratory may render them in harangues and popular addresses, they are certainly, in all discourses that pretend to inform or instruct, wholly to be avoided; and where truth and knowledge are concerned, cannot but be thought a great fault, either of language or person that makes use of them.

Thomas Hobbes (1656) also inveighed against metaphor, characterizing it as an obstacle to genuine communication, a source of ambiguity and obscurity, and thus, a feature of language to be eliminated from philosophical and scientific discourse. It was Giambattista Vico (Bergin and Fisch 1984) who kindled interest in metaphor as evidence of how "knowledge originates from sense," as St. Thomas had so aptly put it. Vico characterized this sense-making capacity as *poetic logic*. Like Aristotle, he saw metaphor as a strategy for explicating or exemplifying abstract notions. However, he went further in claiming that the strategy itself resulted from an association of sense between what is unknown and what is familiar: "It is another property of the human mind that whenever men can form no idea of distant and unknown things, they judge them by what is familiar and at hand" (Bergin and Fisch 1984: 122). The association is not just a matter of convenience or expedience. Rather, the two parts of the metaphor suggest each other phenomenologically. In effect, by saying that *life is a stage*, we are also implying that *stages are life*. The two parts are hardly combined through a mere act of substitution or comparison. A little later, Immanuel Kant (2011, originally 1781) claimed, like Vico, that figurative language was evidence of how the mind attempts to understand unfamiliar things. It was Friedrich Nietzsche (1873) who saw metaphor as humanity's greatest flaw, because of its subliminal power to persuade people into believing it on its own terms. Nietzsche divided human thought into two realms—the realm of perception, consisting of impressions and sensations, and the realm of conception, consisting of the ideas that the mind makes from perception. But conception, Nietzsche asserted, is not a straightforward logical process, but rather, the result of linking impressions together. This linkage is imprinted in the structure of metaphor which, subsequently, has the effect of distorting the true perception of things, creating belief because it prods the mind into perceiving a resemblance among disparate things by simply linking them together in linguistic form. Metaphor is thus the source of our superstitions and of our religious belief systems. In effect, Nietzsche saw metaphor as a linguistic self-fulfilling prophecy. Any attempt to create a universal system of knowledge based on language would be a totally vacuous enterprise because it would be no more than the product of our metaphors.

We can, of course, use metaphor for deception (and certainly do so), but its ubiquity in language and its sapient functions in riddles constitute evidence that the human mind is a connective organ. The fact that getting to the A (Answer) in a riddle (Q) via metaphor is actually a model of how human cognition arguably gains knowledge.

Modern-day interest in metaphor as a feature of mind, rather than as a mere figure of speech, is not due to a reappraisal of Vico or Nietzsche, but rather to the pivotal work of the early experimental psychologists in the latter part of the nineteenth century. One of the founders of the new discipline—the German linguist-physiologist Wilhelm Wundt (1901)—was the first to conduct experiments on how people process figurative language. Another early linguist-psychologist, Karl Bühler (1908), collected intriguing data on how subjects paraphrased and recalled proverbs. He found that the recall of a given proverb was excellent if it was linked to a second proverb; otherwise the proverb was easily forgotten. Bühler concluded that metaphorical-connective thinking produced an effective retrieval form of memory and was, therefore, something to be investigated further by the fledgling science of psychology.

Shortly after Bühler's fascinating work, the Gestalt movement within psychology emerged to make the study of metaphor a primary target of research. Solomon Asch (1950), for instance, examined metaphors of sensation (*hot*, *cold*, *heavy*, and so on) in several unrelated languages as descriptors of emotional states. He found that *hot* stood for *rage* in Hebrew, *enthusiasm* in Chinese, *sexual arousal* in Thai, and *energy* in Hausa (a language spoken in northern Nigeria, Niger, and adjacent areas). This suggested to him that, while the specific emotion implicated varied from language to language, the metaphorical process did not. Simply put, people seemed to think of emotions in terms of physical sensations and expressed them as such. As Roger W. Brown (1958: 146) commented shortly after the publication of Asch's findings, there is "an undoubted kinship of meanings" in different languages that "seem to involve activity and emotional arousal;" and that this "kinship" is revealed through metaphor.

Actually, the scholar who most kindled a broad scientific interest in metaphor was literary critic I. A. Richards. In his groundbreaking 1936 book, *The Philosophy of Rhetoric*, Richards started a veritable revolution in psychology and anthropology by arguing persuasively that metaphor could be described scientifically. The meaning inherent in *life is a stage* is a categorical "interaction" between *life* and *stages*, which today is called a "blend" of specific areas in the brain. As a case-in-point of how Richard's interactionist model would explain the origination of a common metaphor, consider "John is a gorilla." The person named *John* and the animal known as a *gorilla* from a blend in the mind, so that the linkage generates an image of a person with gorilla-like characteristics and, vice versa, of a gorilla with human-like qualities. But this blending would not occur if the two images were not perceived as exemplars of each other in the first place (Fauconnier and Turner 2002). This suggests a different, more general, level of abstraction—one in which animals are perceived as exemplars of human personality and humans as types of animal. If one were to label *John* as a *snake*, a *pig*, or a *puppy*, rather than a *gorilla*, then our image of *John* would change in kind with each new vehicle—he would become serpentine, swine-like, and puppy-like in our imagination. Like Franz Kafka's horrifying short story *Metamorphosis*, where the main character awakes one morning from a troubled dream to find himself changed in his bed to some kind of monstrous creature, our perception of people (and of ourselves for that matter) is altered (probably permanently) the instant we paint a metaphorical picture of their personalities in animal terms. The reason for this is that we perceive ourselves to be animals. Talking about people as if they were animals is a manifestation of this unconscious perception. This is perhaps why many early riddles are about animals as metaphors of human personality. Riddles are also about nature as interconnected with human life, thus bringing out our tendency to "animate the inanimate realm," and to "humanize the animal realm" as a means to understand these realms.

A 1977 study showed that metaphor pervades common speech. Titled *Psychology and the Poetics of Growth: Figurative Language in Psychology, Psychotherapy, and Education*, the study—conducted by a team of psychologists headed by Howard Pollio—it found that speakers of English uttered, on average, an astounding 3,000 novel metaphors and 7,000 idioms per week on average. In other words, their research made it obvious that metaphor could hardly be construed as a deviation from literal language, or a mere stylistic option to it. The study was followed by the 1979 collection of studies by Andrew Ortony, *Metaphor and Thought*, the 1980 anthology put together by Richard P. Honeck and Robert R. Hoffman, *Cognition and Figurative Language*, and the widely read 1980 book by George Lakoff and Mark Johnson, *Metaphors We Live By*. These set the groundwork for a new approach to the study of language that came to be known as *Conceptual Metaphor Theory* (CMT). The central notion on which CMT is implanted is that metaphorical meaning pervades language and thought, and is instinctual, not exceptional. If contextual information is missing from an utterance such as "The murderer is an animal," our inclination is to interpret it metaphorically, not literally. It is only if we are told that the *murderer* is an actual "animal" (a bear, a cougar, and so on) that a literal interpretation comes into focus.

The work in CMT has obvious implications for the study of riddles. It is useful, therefore, to synthesize its formulation in *Metaphors We Live By*. First, Lakoff and Johnson assert what Aristotle claimed two millennia before, namely, that there are two types of concepts—concrete and abstract. But the two scholars added a remarkable twist to the Aristotelian dichotomy—abstract concepts should not be viewed as being autonomous from concrete ones, but rather as blends. They thus renamed an abstract concept a *conceptual metaphor*. To grasp what this designates, recall the example above of *John* and the animals to which he was linked metaphorically (*gorilla, snake, pig, puppy*). When looked at from the perspective of CMT, it is obvious that each specific metaphorical association ("John is a gorilla," "John is a snake," etc.) is an instantiation of an abstract concept, which can be formalized as: *people are animals*. This is an example of what Lakoff and Johnson designated a conceptual metaphor. In effect, metaphors display what can be called "palindromic structure," that is, they can be read both ways conceptually, with the two domains blending into one overarching concept (Danesi 2017). So, in the Riddle of the Sphinx, phases of the day implicate phases of life and vice versa. The two domains constitute a conceptual blend.

Riddles and the origins of culture

Riddles can be divided, generally, into three functional categories—that is, according to their social functions. One of these is to shed light on some aspect of life—a function of the early mythical riddles. Such riddles can be called *mythical*. Riddles, such as the one by Claret, which have a different function, namely diversion and laughter, can be called *ludic*. A third category can be labeled *literary*, defined as the art of riddling as a literary genre.

Mythic riddles connect abstractions, such as phases of life, with concrete source domains, such as phases of a day. The Riddle of the Sphinx falls into this category. If one looks carefully at the themes of the early riddles, it is easy to detect their conceptual metaphorical structure—*life is a journey, humans are animals, objects are divine artifacts*, and so on. Such riddles continue to reverberate latently today as templates for judging human actions and offering advice, bestowing upon everyday life a kind of implicit metaphysical value.

Ludic riddles have recreational and educational functions, constituting intellectual challenges intended to entertain people, much like quiz shows today. They also incorporate conceptual metaphors, but they may also involve a play on words, as we have seen. Derived from this basic ludic function are literary riddle traditions and practices. The riddles of Voltaire or Lewis Carroll, for instance, fall into this category. The mythic function coincides with the cosmogonic and foundational myths of a culture, embedded within them (as the Riddle of the Sphinx). These arise from *mythos*, or dreaming and unconscious culture. The ludic and literary riddles arise, instead, from *lógos*, that is, they are designed consciously as intellectual or recreational activities. The shift from unconscious (*mythos*) to conscious (*lógos*) culture appears to be a principle of human evolution—a process that repeats itself in cycles.

The study of mythic riddles suggests that culture is born in *mythos*. In a way, these riddles are the language of dreams. As a culture evolves, it is guided by *lógos*; it is at this point that ludic riddles emerge—that is, riddles with social-recreational functions. The early mythic riddles have an existentialist function—namely, to articulate interpretations about problems of existence; the ludic riddles have a "comic relief" function, that is, an escape from those problems.

A similar view of culture origins was set forth by the French sociologist Émile Durkheim (1912). Durkheim rejected the notion that myth arises in response to extraordinary manifestations of Nature. To Durkheim, Nature was a model of regularity and thus predictable and ordinary. He concluded that myths arose as emotional responses to existence, thus constituting a narrative existential code and a system of historical reasoning. Myths and the rituals stemming from them sustain and renew the code, keeping it from being forgotten, and binding people socially. He explained the remarkable similarities among the world's myths with his notion of a "collective conscious," by which the basic ingredients of myth are actually part of the human brain and thus common to every human being (Durkheim 1912: 12):

> The collective conscious is the highest form of the psychic life, since it is the consciousness of the consciousness. Being placed outside and above all individual and local contingencies, it sees things in this permanent and essential aspect, which it crystallizes into communicable ideas ... it alone can furnish the mind

with the molds which are applicable to the totality of things and which make it possible to think of them.

The anthropologist Bronislaw Malinowski (1922) also claimed that myth provided a rationale for coming to grips with the problems of existence. He argued that, in its primitive form, myth was not merely a story, but a lived emotional reality. It was not fictional, and it lives on in our rituals, governing our modes of perception and controlling our conduct in an unconscious fashion. In myth, the identity and basic values of the group were thus given an absolute meaning. The most popularized studies of myths in the twentieth century were those of the American scholar Joseph Campbell (for example, 1972). Campbell combined insights from Jungian psychology and linguistics to formulate a general theory of the origin, development, and unity of all human cultures. If there is thunder in the sky, and one lacks the notion of "thunder," then it can be explained as the angry voice of a god; if there is rain, then it can be explained as the weeping of the gods; and so on. A myth is a telling of such events. It is relevant to note that the same themes of the myths are embedded in the early riddles—the two are often co-occurrent historically.

This view of the ancient mythic riddles as expressions of early unconscious archetypal forms of culture is corroborated, indirectly, by the universality of the themes in ancient riddles across the world. The *phases of life* riddle is found virtually across ancient cultures. An example is found in an Estonian myth (Aarne 1918: 12):

It goes in the morning on four feet,
at lunch-time on two,
at evening on three.

The *year unfolds like a wheel* riddle is found across Europe and Asia. A Sanskrit riddle, for instance, describes a "twelve-spoked wheel, on which stand 720 sons of one birth" alluding to the twelve months of the year, which together have 360 days and 360 nights (Tupper 1903: 102). Animal riddles are common not only in ancient Greece or Rome but are found virtually everywhere. The following early medieval Icelandic riddle is a metaphorical portrayal of a cow, which has four udders, four legs, two horns, two back legs, and one tail (Aarne 1918: 60):

Four are hanging,
Four are walking,
Two point the way out,
Two ward the dogs off,
One ever dirty
Dangles behind it.

This riddle ponder
O prince Heidrek!

This line of argument may explain why riddles were seen as the language of oracles, who would discuss the meaning of dreams in riddle form. Riddles are evidence of the human brain's ability to visualize the universe as a coherent organism. They emerge as oral tradition and as intrinsic to the expression of wisdom. Throughout cultures, they were perceived intuitively as being timeless and revealing universal truths. It was to gain relief from the contemplation of truth that ludic riddles surfaced in cultural life. In early mythic riddles, there is an unconscious need to search for the origin of things; in the later ones, a need to laugh at ourselves for searching in the first place. This existential dualism continues to beset the modern imagination.

3

Word Games

A character is like an acrostic or Alexandrian stanza; read it forward, backward, or across, it still spells the same thing.

Ralph Waldo Emerson (1803–82)

Riddles, as discussed, possess the Q → (A) dialectical structure, constituting open texts (Chapter 1) in which the answer must be uncovered by ingenuity. On the other hand, games are closed texts, because we are given the end-state in advance; the challenge is to devise a way to get to the end-state under given stipulations or rules. In a scrambled Rubik's Cube, for instance, we are not asked to search for a hidden answer, but rather to attain an end-state. Many common word games have the same structure—they ask us to reach an end-state or outcome via certain rules or conditions. Included in this category are anagrams and acrostics. Consider a typical anagram:

How many words can be made with "live" by rearranging all its letters (answer: veil, evil, vile)?

Of course, we do not know what the anagrammatized words are; but we know that they are there for us to identify. As mentioned (Chapter 1), the terms *puzzle* and *game* are used virtually synonymously today, but as this example shows, there is a cognitive difference. Whatever we call them, games based on language structure abound throughout cultures and across time, and some of them are intrinsically intertwined with the myths, legends, and traditions of the ancient societies. Name anagrams, for instance, were widely perceived to contain secret messages, predicting or affecting the destiny of the name-bearer. It was only after the Renaissance that the widespread view of names as "magical anagrams" started to fade, as society began reinterpreting the symbols of the past in new scientific ways. But the feeling that words possess magical qualities was not erased from the human imagination. This feeling can still be seen on children's faces each time they learn how to play a word game successfully.

This chapter deals with word games. Needless to say, it is impossible to deal here with all types and genres of word games across cultures. So, the discussion here will be illustrative and limited to the most generic types. An in-depth treatment of word games by Eckler (1997) shows how truly vast this domain of enigmatology is.

Anagrams and acrostics

There are four main kinds of anagrams: (1) word-to-word anagrams; (2) word-to-phrase anagrams; (3) phrase-to-word anagrams; and (4) phrase-to-phrase anagrams. In the case of (1), the challenge is to rearrange the letters of a given word to produce a new word. For example, the letters in the word *stop* can be rearranged to produce *pots*, *opts*, *tops*, *spot*, or *post*. For type (2), the idea is to produce a phrase with the letters of a given word: for example, *dormitory* will yield *dirty room*. Type (3) is the reverse of the previous one. The idea is to produce a word from a phrase: for example, the letters in the phrase *Is pity love*? can be rearranged to form the single word *Positively*. Finally, the objective in type (4) is to rearrange the letters of a given phrase to form another phrase (or sentence). For example, rearranging the letters of the phrase *the countryside* will yield *No city dust here*.

The origin of anagrams as a divinatory art is shrouded in mystery, which appears around the time of Moses, as the craft of decoding the hidden meaning of personal names (Wheatley 1862: 72). The Hebrews from the Temple of Rehovot are often credited with its invention, probably because the Cabbalists were fond of the art. Anagrams surface in Greek and Roman societies, although they are nearly all lost. The belief in the mystical power of anagrams to reveal truth likely is a consequence of the origins of writing as a sacred form of representation. The ancient Egyptians, in fact, called their writing system *hieroglyphic* (*hieros* "holy" and *glyphein* "to carve") because it was used to record hymns and prayers, to register the names and titles of deities, and to record various sacred activities. All ancient scripts were thought to have sacred origins—for example, the Cretans attributed the origin of writing to Zeus, the Sumerians to Nabu, the Egyptians to Toth, the Greeks to Hermes, and so on.

The earliest form of writing was pictographic, realized by drawing representative pictures to stand for objects directly. One of the first civilizations to institutionalize pictographic writing as a means of recording ideas was the ancient Chinese one. According to some archeological estimates, Chinese writing may date as far back as the fifteenth century BCE. Another fully developed ancient pictographic system was the Sumerian-Babylonian one that emerged 5,000 years ago. The Sumerians recorded their writings on clay tablets with wedge-shaped forms, hence the name cuneiform. This was a very expensive and impracticable means of writing. For this reason it was developed, learned, and used primarily by rulers, nobles, and clerics. In Egypt, hieroglyphic writing emerged around 2700–2500 BCE. The Egyptians eventually used papyrus (a type of early paper made from reeds) on which to record their writings, making it more practicable for many more classes of people. To facilitate the speed of writing, the Sumerians and the Egyptians eventually streamlined their pictographs, transforming them into symbols for sounds of speech. Called phonographic writing, it consisted of signs standing for parts of words, such as syllables or individual sounds. A complete system for representing speech sounds is called alphabetic. The first such system emerged in the Middle East around 1000 BCE, and was then transported by the Phoenicians to Greece. It contained signs for consonant sounds only. In Greece, signs for vowel sounds were added to it, making it the first fully fledged alphabet.

With the development of alphabets, the practice of name anagrams became quite common. In the third century BCE, the Greek poet Lycophron (285–247 BCE) devised anagrams on the names of members of the court of Hellenistic King Ptolemy II. His two most famous anagrams were on the names of Ptolemy and his queen, Arsinoë, in his famous poem on the siege of Troy titled *Cassandra*. Anagrams were hardly perceived as mere games in the ludic sense—they were perceived as having the same kind of prophetic power as riddles. Emperors and socially important people even hired anagrammatists to foretell their destinies, much like we would consult fortune-tellers today. Legend has it that Alexander the Great firmly believed in their prophetic power. During the siege of the city of Tyre, he was particularly troubled by a dream he had in which a satyr appeared to him. The next morning he summoned his annagramatists to interpret the dream. They pointed out to Alexander that the word "satyr" itself contained the answer, because in Greek "satyr" was an anagram of "Tyre is thine." Reassured, Alexander went on to conquer the city on the subsequent day.

One of the most famous anagrams of all time was constructed in the Middle Ages, according to Samuel Johnson (1755). Its unknown author, Johnson asserted, contrived it as a Latin dialogue between Pilate and Jesus. Jesus' answer to Pilate's question "What is truth?" is phrased as an ingenious anagram of the letters of that very question (Hovanec 1978: 67):

Pilate: Quid est veritas? ("What is truth?")
Jesus: Est vir qui adest ("It is the man before you")

Anagrams had become vastly popular in the medieval era, when the right to critique or ridicule royalty and politicians with anagrams was enshrined in English law in the *Magna Carta* (1215), whose letters can curiously be rearranged to *Anagram Act*. Interestingly, philosophers and scientists of the seventeenth century hid their discoveries in the form of Latin anagrams, allowing them to lay claim to them by concealing them in this way, before their results were ready for publication.

The belief that anagrams were the linguistic vehicles for predicting destiny lasted until the late Renaissance. Soothsayer status was often earned by those who showed that they possessed a deep knowledge of anagrams, especially those who knew how to foretell the fates of individuals through the anagrams of their names. People would often wear amulets with anagrams of their names on them to ward off evil. Nowhere is the belief that letters hide fateful messages more evident than in the tradition of using *runes*—the ancient alphabetic characters of Germanic peoples—to predict the future. In elaborate rituals, tokens bearing the runes were put into a container and shaken. Placing one's hand into the container was said to produce an urge to select one rune in particular. The rune drawn was then thought to predict the individual's fate. For instance, the V symbol (*vara*) was a portent of healing (*Vara* was a Viking goddess who escorted the souls of warriors to Valhalla.).

Anagrams have even been constructed retrospectively to explain a person's fate. The classic example concerns Mary Queen of Scots, who died by execution, and was posthumously memorialized with the Latin expression *Trusa vi regnis, morte amara cado* ("Thrust by force from my kingdom I fall by a foul death"), which is an anagram of *Maria Steuarda Scotorum Regina* ("Mary Stewart Queen of Scots") (D'Israeli 1835: 183). Another example relates to Henry IV of France, who was assassinated in 1610 by an unscrupulous man named Ravillac. After Henry's death, it was pointed out, ominously, that *Henricus Galliarum Rex* ("Henry King of the Gauls"), when rearranged, became *In herum exurgis Ravillac* ("From these Ravillac rises up") (Wheatley 1862: 79). The move away from the divinatory function of anagrams to ludic functions, likely started in seventeenth-century France, when Louis XIII employed a "Royal Anagrammatist" to prepare challenging anagrams for him for his own delectation. The post came with a substantial salary and a rank as high as any assigned to the famous poet laureates of the era. Louis, however, also indulged himself by getting his Anagrammatist to rearrange the names of famous people, in order to glean from this any insights into their character. A widely cited anagram of the latter type is by Lewis Carroll. He rearranged the letters of the name of the British humanitarian, Florence Nightingale, resulting with a fitting eulogy (cited by Costello 1988: 38):

Florence Nightingale = Flit on, cheering angel!

Although we might now consider it a coincidence or a curiosity that name anagrams seem somehow connected to an individual's life or personality, we still harbor within us an instinctive belief that they do indeed reveal "truth," in the same cautious and half-hearted way that we believe horoscopes do. An anagram of Clint Eastwood's name, which describes his early career, is rather prophetically "Old West action." Below are a few well-known examples of this kind:

Alec Guinness = Genuine class
Princess Diana = End is a car spin
Henry Wadsworth Longfellow = Won half the New World's glory
England's Queen Victoria = Governs a nice, quiet land

Words formed by taking the first letter of each line of a verse or group of words, are called *acrostics* (from the Greek *akron* "head" and *stikhos* "row, line of verse"). The first four poems of the *Book of Lamentations* are acrostics. The first, second, and fourth, which are patterned after older funeral songs, comprise twenty-two verses, beginning with a successive letter of the Hebrew alphabet; the third is an individual lament, containing sixty-six verses, with three verses to each letter of the alphabet. The fifth verse is a group lament, again containing twenty-two verses, but it is not acrostic. All five laments speak movingly of God's harsh chastisement of his people for their sins. Coupled with these anguished verses are lines recalling God's mercifulness and expressing the prayerful hope that he may abate his wrath and restore a chastened Israel. The first part of the Book of Nahum (1: 2–11) is an unfinished acrostic poem, constructed with roughly half the letters of the Hebrew alphabet. The poem depicts God as jealous and angry, ready to take vengeance on those who oppose him. In the eighth section of the Book of Proverbs, a collection of short moral sayings, yet another twenty-two-line acrostic poem can be found. And in Psalm 119, each eight-line stanza similarly begins with a successive letter of the Hebrew alphabet. Known appropriately as the "Abecedarian Psalm," it is the oldest acrostic known.

Acrostics, like anagrams, are embedded in *mythos*. The most famous one in this regard is the so-called *Sator Acrostic*, found in the Roman city of Cirencester in England, on a column in the city of Pompeii, and in other ancient sites (Atkinson 1951), shown in Figure 3.1.

R	O	T	A	S
O	P	E	R	A
T	E	N	E	T
A	R	E	P	O
S	A	T	O	R

Figure 3.1 The Sator acrostic.

The word *ROTAS* starts in the top leftmost cell and can be read across the top row and vertically down the leftmost column; the same word starts, palindromically, in the bottom rightmost cell and can be read from right to left across the bottom row and upward in the rightmost column. The word *OPERA* starts in the second square of the top row, and can be read downward in the second column; it can also be found starting in the second square from the top in the leftmost column, from where it can be read from left to right across the second row. Moreover, it appears palindromically in the opposite squares: that is, it starts in the second-to-last square from the bottom in the rightmost column, from where it can be read from right to left; and it starts in the second-to-last square from the right in the bottom row, from where it can be read upward.

All the words in the square follow this "palindromic mirror pattern." A generally accepted translation of this acrostic is: "Arepo, the sower, carefully guides the wheels," which is construed by some scholars as a metaphor for "God controls the universe" (Atkinson 1951; Bombaugh 1962: 348). Its mystical-sacred meanings is the reason why, throughout the medieval period, it was commonly carved into amulets to ward off disease (Holroyd and Powell 1991: 78). But the word "Arepo" argues somewhat against this Christian interpretation because it appears to be of Celtic origin, meaning "plough." Moreover, since it has been found incised on a column in Pompeii, the only way to ascribe a Christian origin to the acrostic is to hypothesize that it was added to the column by Christians of a later period. As a sacred artifact, the letters of the acrostic were anagrammatized in the medieval period in the form of a cross to spell out *Pater Noster*, "Our Father." The two *A*s and *O*s on the arms of the cross also allude to the *alpha* and the *omega*, the first and last letters of the Greek alphabet and thus, metaphorically, "the beginning" and "the end."

The reputed inventor of palindromes was Sotades, a minor Alexandrian poet of the early third century BCE of whom very little is known. As Bombaugh (1961) has amply illustrated in his collection of word games, palindromes are found throughout history and across languages. Many were apparently constructed on purpose for socially significant events. Actually, the Sator Acrostic constitutes a particular type of word square, a grid of letters which conceals words or messages. One of the earliest examples, dating from the second or third century CE, is a thirty-nine by thirty-nine square array of Greek letters, carved in alabaster by an Egyptian sculptor known as Moschion. To read the square one must start at the center and read right or left, up or down, turning at right angles along the way. This reveals the phrase "Moschion to Osiris, for the treatment which cured his foot," which is repeated over and over, in everlasting tribute to the healing god.

Acrostics migrated to the domain of recreational game playing in the nineteenth century, when Queen Victoria made ingenious acrostics, such as the one below, which dates to 1856. The initial letters in the answer column, read downward, spell out the name of an English town, and the final letters, read upward, tell what that town is famous for (Costello 1988: 13):

A city in Italy
A river in Germany
A town in the US
A town in North America
A town in Holland
The Turkish name of Constantinople
A town in Bothnia
A city in Greece
A circle on the globe

The answers are *Naples*, *Elbe*, *Washington*, *Cincinnati*, *Amsterdam*, *Stamboul* (antiquated name for Istanbul), *Tornea*, *Lepanto*, and *Ecliptic*. Now, taking the first letters in each of these answers produces: *Newcastle* which is famous for its *coalmines*. The latter word is constructed with the last letters of the answers in reverse order. Lewis Carroll wrote scores of acrostic poems in the same century, raising them to a literary genre. The final poem in *Through the Looking-Glass* conceals the name of the young girl on whom the character of Alice was based.

Like riddles, anagrams and acrostics migrated gradually to the ludic and literary domains, that is, from *mythos* to *lógos*. Nowadays, in fact, anagrams and acrostics are positioned conceptually as simple word games. Nevertheless, anagrams are still employed as portentous messages by pop culture and writers today. A famous example is in Roman Polanski's 1968 horror masterpiece *Rosemary's Baby*. Three weeks before her delivery date, Rosemary gets a call informing her that Hutch has tragically died. At the burial ceremony, she is given a book titled *All of Them Witches*, which dealt with the practices of witches and warlocks. One of the warlocks she reads about is named Adrian Marcato. She notices, to her horror, that the name of Adrian Marcato's son Steven is underlined, whereupon she realizes that Steven Marcato is an anagram for Roman Castevet, her neighbor. Rosemary thus starts to realize that the Castevets and their friends are members of a coven, and that her unborn baby's blood will be used in one of their rituals.

Perhaps the ancients were right all along in claiming that anagrams were reifications of hidden messages, as we confront today one of the most difficult anagrams ever devised by nature—DNA. Composed of sequences of chemical compounds called nucleotides, DNA is the genetic material found within the cells of all living things. With the exception of identical siblings, all individuals have unique DNA. In a fundamental biological sense, therefore, the nucleotides that make up an individual's DNA are anagrams that do indeed predict the individual's destiny.

Cryptograms

Cryptography is the science of encoding and decoding messages. It originated in antiquity in the military world. During the Peloponnesian War, in the fifth century BCE, Spartan soldiers communicated with their field generals during battle by writing messages across a strip of parchment wrapped spirally around a staff called a *scytale*. When the parchment was unwound, the message became unreadable—except to the generals, who had another staff of the correct thickness and could rewrap the parchment around it. No one seems to have noticed the importance of cryptography to the philosophy of mind or to mathematics, until centuries later. The first modern work on the mathematics of cryptography was written, in fact, by a German abbot named Johannes Trithemius in 1510.

Today, we perceive cryptography as connected more to mystery than to mathematics. The 2004 movie, *National Treasure*, for example, revolved around cryptography. An encrypted map to a lost treasure is decoded by protagonist Benjamin Franklin Gates, an historian and amateur cryptanalyst, whose name is an obvious allusion to Benjamin Franklin, an expert amateur cryptographer himself. The map is found on the back of the Declaration of Independence. It is quickly determined that it will lead to the greatest treasure in American history, recalling previous classic stories of treasure-hunting based on cryptography, such as Edgar Allan Poe's "The Gold Bug" (1843). The encrypted message is linked to the "Silence Dogwood" letters—letters written by Benjamin Franklin under this pen name, providing the insight for cracking the cipher, which leads the investigators to the bell tower of Independence Hall. There they find a pair of glasses. When used to read the back of the Declaration, the glasses reveal a critical clue: "Here at the Wall." From subsequent clues the treasure is eventually located. The appeal of the plot is undoubtedly shaped by cryptography—decode the map and you will unravel a mystery.

The first to use cryptography as part of a fictional story was, as mentioned, Edgar Allan Poe, who was also the inventor of the detective genre with two stories—"The Murders in the Rue Morgue" (1841) and "The Purloined Letter" (1844). Poe's detective hero figure, C. Auguste Dupin, solved crimes by interpreting the clues left behind by the criminal. The kind of thinking used by Dupin is an example of what Charles Peirce called *abduction* (Peirce 1938). This is an inference-based form of reasoning, complementing deduction and induction. Dupin used all three modes at various stages of his investigations. He applied deductive reasoning to classifying the clues in a systematic way; he employed inductive reasoning to draw a general picture of the situation; and he used abduction to both envision and then interpret that picture, which ultimately reveals the true story behind the crime. Actually, this model of crime-solving fits perfectly with the one proposed in this book for solving all kinds of puzzles. All three modes may be used to various degrees, but abduction seems to be particularly prominent for figuring out what a puzzle involves. So, the model put forth in the opening chapter can be modified somewhat as follows:

Imagination → hunch (abduction) → visualization (induction) → Aha insight → solution (deduction)

As Peirce (1931–58, Vol. 7: 231) observed, abduction is a powerful feature of the human mind, even though it may appear to be no more than guessing; it reveals how we make sense of things:

> No new truth can come from induction or from deduction. It can only come from abduction; and abduction is, after all, nothing but guessing. We are therefore bound to hope that, although the possible explanations of our facts may be strictly innumerable, yet our mind will be able, in some finite number of guesses, to guess the sole true explanation of them. That we are bound to assume, independently of any evidence that it is true. Animated by that hope, we are to proceed to the construction of a hypothesis.

Poe's "The Gold Bug" revolves around a cipher, supposedly devised by the pirate Captain Kidd. The narrative starts with the protagonist, Legrand, coming into possession of a bug. His servant, Jupiter, is convinced that the bug is made of pure gold, and fears that his master might have been bitten by it, rendering him insane as a result. Without going into the details of the narrative, the main aspect of the story is a cryptogram, wrapped in a piece of parchment, in which Legrand noticed traces of invisible ink, brought out by the effects of the heat of a fire burning on his hearth. The parchment reveals a hidden cryptogram, which Legrand deciphers—it contained directions for finding a treasure buried by Captain Kidd. The cryptogram, as displayed in the story, is shown in Figure 3.2.

```
53 ‡ ‡ † 305)) 6*; 4826) 4 ‡ . ) 4 ‡); 806*; 48 † 8
¶ 60 )) 85; ; ] 8* ; : ‡ *8 † 83 ( 88 ) 5* † ; 46 ( ; 88*96
* ? ; 8 ) * ‡ ( ; 485 ) ; 5* † 2 : * ‡ ( ; 4956 * 2 ( 5*–4 ) 8
¶ 8 * ; 4069285 ) ; ) 6 † 8 ) 4 ‡ ‡ ; 1 ( ‡9 ; 48081 ; 8 : 8 ‡
1 ; 48 † 85 ; 4 ) 485 † 528806*81 ( ‡9 ; 48 ; ( 88 ; 4
( ‡ ? 34 ; 48 ) 4 ‡ ; 161 ; : 188 ; ‡ ? ;
```

Figure 3.2 Cipher in "The Gold Bug" (from Wikimedia Commons).

This is a Polybius cipher, with numbers and symbols standing for letters. Adding word spaces, punctuation, and capitalization the hidden message (directions for finding the treasure) is as follows:

> A good glass in the bishop's hostel in the devil's seat twenty-one degrees and thirteen minutes northeast and by north main branch seventh limb east side shoot from the left eye of the death's-head a bee line from the tree through the shot fifty feet out.

Legrand inferred that the "bishop's hostel" indicated the site of an ancient manor house, where he found a narrow ledge that resembled the outline of a chair (the "devil's set"). Using a telescope with the given bearing, Legrand spotted a skull among the branches of a large tree through which a weight had to be dropped—the bug itself—from the left eye in order to locate the treasure, which subsequently happens successfully.

"The Gold Bug" was an instant success, becoming Poe's most widely read work during his lifetime. It also helped popularize cryptograms as puzzles in themselves. Like other literary figures of his time, Poe had developed a keen interest in cryptography, believing that it stimulated the use of the imagination, together with reason and logic. In the December 1839 issue of *Alexander's Weekly Messenger*, he egged readers on to write to him using ciphers rather than typical letters. Poe had realized that his readers shared his enthusiasm, which induced him to write "The Gold Bug." Interestingly, Poe's story also stimulated scientific and professional interest in cryptography, above and beyond previous scholarship. It inspired William F. Friedman, a famous military cryptanalyst, who read "The Gold Bug" as a child. Freidman became renowned for cracking Japan's so-called PURPLE code during World War II (Clark 1977: 103–12).

Inspired by Poe, Sir Arthur Conan Doyle incorporated cryptography into three of his Sherlock Holmes stories (Aliseda 2010): "The Adventure of the *Gloria Scott*" (1893), where a message is concealed within a text as every third word; "The Valley of Fear" (1915), in which a book code is featured; and "The Adventure of the Dancing Men" (1903). The latter story involves a code based on twenty-six stick men figures, each one representing a letter of the alphabet, around which the entire narrative revolves. The cryptic message is introduced when Mr. Hilton Cubitt of Ridling Thorpe Manor in Norfolk goes to consult Sherlock Holmes, giving him a piece of paper with a mysterious sequence of stick figures (Figure 3.3).

Figure 3.3 Stick figures in "The Adventure of the Dancing Men" (from Wikimedia Commons).

The code is a simple substitution cipher, with each figure standing for a specific letter of the alphabet. The meaning of various ciphers in the story concerns a past secret, which comes back to haunt Elsie, Cubitt's wife. The secret is a romantic involvement with an American gangster. As it turns out, by keeping her past concealed, she ends up losing her husband and is condemned to live without him for the rest of her life.

Cryptography has appeared frequently in the detective-crime-mystery genre since its introduction by Poe and Conan Doyle. John F. Dooley (2016) has identified nearly 400 fictional works that include secret writing. But even before the detective genre, cryptographic themes and elements are found scattered in fiction. William Makepeace Thackeray's historical novel, *The History of Henry Esmond, A Colonel in the Service of Queen Anne, Written by Himself* (1852), revolves around the life of Henry Esmond, Viscount of Castlewood, who became involved in a plot to ensure the succession of James III as King of England. Early in the novel, there is mention of cipher messages connected with Henry's uncle's support of King James II during the Glorious Revolution in 1688. The more interesting of the ciphers uses a Cardano grille to hide a message in a letter from Henry's cousin Frank to his mother detailing the arrival of the Pretender James. Neither the cipher nor the key is described in the novel, leaving the reader in a state of unease. The Cardano cipher reverberates with historical significance, giving the novel a context in which mystery and political mayhem coincide. In 1550, Italian mathematician and astrologer Girolamo Cardano invented a simple grid for writing hidden messages, which would conceal a message within the layout of an ordinary letter so that the whole text would not appear to be a cipher at all. It was Cardinal Richelieu who used the grille for both private and diplomatic correspondence, leading to its general use in letters and literature in the seventeenth century—the century in which part of Thackeray's novel takes place.

Jules Verne incorporated cryptography inventively in several of his novels. Perhaps the best known one is *La Jangada* (1881), translated in English as *Eight Hundred Leagues in the Amazon*. The main character, Judge Jarriquez, deciphers a substitution cipher, saving an innocent man from the gallows. Verne describes the thought processes used by Judge in a detailed manner, as he struggles to decode the message, thus constituting a kind of cryptanalytic lesson. Verne followed this story with *Mathias Sandorf* (1888). The hero, Mathias Sandorf, enciphers a message using a rotating six by six Cardano grille, which is intercepted by the villains in the story who decipher the message. Verne gives a good description of how they cracked the message without providing the message itself, leaving the reader to engage in both the mystery plot and in solving the mystery cipher, thus blending them seamlessly into an early example of an "interactive" narrative (Singh 1999).

Following on Verne's coattails was Maurice Leblanc with his fascinating story, *The Hollow Needle*, published in 1909. The protagonist is a thief named Arsène

Lupin—an allusion to Poe's Dupin—who crosses paths with the famous Inspector Holmlock Shears—a quasi-anagram of Sherlock Holmes. Arsène Lupin is opposed by Isidore Beautrelet, a young gifted amateur detective, who is poised to engage Lupin in a contest of wills. The *Hollow Needle* in the title refers to an actual historical fact—the second secret of Marie Antoinette and the Italian occultist Alessandro Cagliostro concerning the hidden fortune of the Kings of France. The secret is revealed to Lupin by the character Josephine Balsamo, in Leblanc's later novel, *The Countess of Cagliostro* (1924). The Kings of France had been handing down the secret to each other for ages. Lupin decoded it as part of the story—the legendary needle contains regal dowries, pearls, rubies, sapphires, and diamonds.

Agatha Christie, the successor to Poe and Conan Doyle, also used cryptography in several works, including a flower-name code in *The Four Suspects* (1931) solved by Miss Marple. In a similar vein, using cities instead of flowers for her code, Dorothy L. Sayers incorporated a cipher as part of the plot of her *Have His Carcase* (1932). The detective in the story solves the crime by deducing that the key is the name of a city and a year.

Cryptography has played a significant role in various blockbuster movies. One well-known example is *Johnny Mnemonic* (1995), based on the novel by William Gibson. In a world dominated by mega-corporations, automatons, and AI, secrecy is of utmost importance—both to individuals and dissident groups. Johnny Mnemonic has a data storage implant in his brain that allows him to carry information (as a "mnemonic courier") secretively in order to transfer it across the Net (the virtual-reality descendant of the Internet). At the same time, though, it allows him to compute and decode the meaning of events as they occur, mirroring the process involved in cracking a code. The crisis comes when Johnny has to upload data at the risk of his own life. He does so just the same and, as a consequence, the dissident group he represents is massacred although he escapes with his life, having come across the encryption password, which allows him to flee. The subtext is a cautionary tale for the times—the world of technology is depriving us of our humanity, which we must resist before it takes over.

Other works of fiction that employ cryptography include *Digital Fortress* (1998), *Mercury Rising* (1998), *Crytonomicon* (1999), *Enigma* (2001), and *The Imitation Game* (2014), among others. Given the extraordinary feats of ingenuity that have gone into inventing and breaking secret codes in wartime and fiction, it is little wonder that the appeal of cryptography has extended beyond these realms, motivating a genre of puzzles that gained popularity first in the 1950s (Gaines 1956, Gardner 1972). The three most common types involve: (1) replacing the letters in a message with numbers (a *letter-to-number cryptogram)*; (2) replacing the letters with other letters (a *letter-to-letter cryptogram*); and (3) replacing the letters with specific types of symbols (a *letter-to-symbol cryptogram*). The first type, as mentioned, is a Polybius cipher, after the Greek historian Polybius, who was among the first to encrypt messages by converting letters to numbers. An example of such a cryptogram is given below:

In the following ciphertext: (1) 1 = T, 5 = O, and 6 = S, (2) there are no abbreviations, (3) the letter N does not appear. What message does it conceal?

1 2 3 4 5 6 3 7 6 1 2 3
6 8 9 10 5 11 5 12 11 5 13 3

The successful solution to a cryptogram hinges on knowing linguistic facts—for example, what single-letter words are possible in a language, what sequences of letters are the most frequent, what is the normal structure of words, and so on. The first part of the solution strategy is to replace the numbers with the given letters. This produces the following partial plaintext:

T 2 3 4 O S 3 7 S T 2 3
S 8 9 10 O 11 O 12 11 O 13 3

Now, let's consider the first word in terms of common English sentences. It starts with a *T*. At the beginning of a sentence, a common word is the article *THE*. If this choice is incorrect, then the solution will go nowhere. We would then have to start again with another hypothesis. Trial-and-error is an intrinsic part of the solution process. This tentative assumption allows us to establish the following matches: *2* = *H* and *3* = *E*:

T H E 4 O S E 7 S T H E
S 8 9 10 O 11 O 12 11 O 13 E

There are four possibilities for the second-last word in terms of two-letter sequences in English: *OF*, *OK*, *ON*, *OR*. Since, there are no abbreviations and no *N*, then the only two possibilities are *OF* or *OR*. We will try our luck with the former. If this should not work out, then we can always go back to this point and use *OR*. So, *12* = *F*:

T H E 4 O S E 7 S T H E
S 8 9 10 O 11 O F 11 O 13 E

Continuing to reason in this way, the plaintext turns out to be: *THE ROSE IS THE SYMBOL OF LOVE*. The same type of thinking is required to solve any other type of cryptogram, such as the letter-to-letter one, called a *Caesar cipher*. In the following example, for instance, the key linguistic insight is that the letters in the plaintext have been shifted two spaces later in the normal alphabetic sequence:

JGNNQ

J was used for *H, G* for *E, N* for *L*, and *Q* for *O*. Making the appropriate substitutions produces the answer: *JGNNQ* = *HELLO*. Ingenious techniques for constructing Caesar ciphers are found scattered throughout the annals of history. The sacred Jewish writers of ancient times, for instance, concealed their messages by reversing the alphabet, that is, by using the last letter of the alphabet in place of the first, the next-to-last in place of the second, and so on. The most challenging Caesar ciphers are, needless to say, random substitutions.

Solving letter-to-symbol ciphers involves the same type of thinking and need not be discussed here. Rather than using numbers, various other symbols are used. A widely used method of encoding messages is known as *transposition*. This involves transposing or reordering the plaintext letters in various ways. As an example, consider the following plaintext:

ALEX LOVES APPLES.

This can be transposed in several ways: (1) by writing the words backward (*XELA SEVOL SELPPA*); (2) by separating the vowels from the consonants (*AELX OELVS AEPPLS*); (3) by reversing syllables wherever possible (*EXAL ESLOV PLESAP*); and (4) by other transformations. It can also be enciphered by a combination of substitution and transposition methods. For example, the letters of the above message can be assigned number values in the order they appear, and this method can then be applied to every letter. At the same time, the transposition method of writing the words backward can be applied to the text. This would produce an extremely difficult message to unravel indeed.

Recall that one of the ways to differentiate puzzles from games is in terms of an open-versus-closed dichotomy—the former involves discovering the answer itself; the latter entails devising how to get to the given end-state. Cryptograms seem to fall in-between the two. The plaintext is hidden and thus must be figured out in the same way that any hidden answer is; however, a large part of cryptograms is figuring out what method is applicable (substitution, Polybius, and so on) in order to uncover the key that leads to the solution.

Cryptograms are probes of linguistic structure. Even a perfunctory calculation of the letters that start words in a dictionary will suggest the presence of consistent statistical patterns in word-formation. Russian mathematician Andrey Markov was among the first to study this very phenomenon in 1913. He sought to determine whether a specific writer's style was governed by frequency factors. He was particularly intrigued by Pushkin's style, as Barrow (2014: 237–8) observes:

> Markov looked at an extract from Pushkin of 20,000 (Russian) letters which contained the entire first chapter and part of the second chapter of a prose poem, with its characteristic rhyming patterns . . . Markov simplified Pushkin's text by ignoring all punctuation marks and word breaks and looked at the correlations of successive letters according to whether they were vowels (V) or consonants (C). He did this rather laboriously by hand (no computers then!) and totaled 8,638 vowels and 11,362 consonants. Next, he was interested in the transitions between successive letters: investigating the frequencies with which vowels and consonants are adjacent in the patterns VV, VC, CV or CC. He finds 1,104 examples of VV, 7,534 of VC and CV and 3827 of CC. These numbers are interesting because if consonants and vowels had appeared randomly according to their total numbers we ought to have found 3033 of VV, 4755 of VC and CV and 7457 of CC. Not surprisingly, Pushkin does not write at random. The probability VV or CC is very different from VC and this reflects the fact that language is primarily spoken rather than written and adjacent vowels and consonants make for clear vocalization. But Markov could quantify the degree to which Pushkin's writing is non-random and compare its use of vowels and consonants with that of other writers, If Pushkin's text were random then the probability that any letter is a vowel is 8,638/20,000 = 0.43 and that it is a consonant is 11,362/20,000 = 0.57. If successive letters are randomly placed then the probability sequence VV being found would be 0.43 × 0.43 = 0.185 and so 19,999 pairs of letters would contain 19,999 × 0.185 = 3,720 pairs. Pushkin's text contained only 1,104. The probability of CC is 0.57 × 0.57 = 0.325. And the probability of a sequence consisting of one vowel and one consonant, CV or VC, is 2 (0.43 × 0.57) = 0.490.

Leaving aside the fact that the results might be specific to the Russian language, and more specifically to Pushkin's own style, Markov's analysis is still a remarkable one. It implies that factors such as discourse genre, semantics, and other linguistic factors have a direct influence on the form and structure of a text and that this can be determined statistically. Markov's approach eventually developed into the field of *corpus linguistics* or the study of linguistic corpora in order to extrapolate from them patterns in actual linguistic usage. Markov's approach, actually, has been critical to modern-day cryptanalysis. For example, an algorithm might be devised to match letters with word sequences and given instructions for identifying homophones and synonyms. The key cryptographic insight is that the letters of the alphabet are not equally distributed in actual

Table 3.1 Letter frequencies in English

Letter	Average number of occurrences per 100 characters	Letter	Average number of occurrences per 100 characters
E	12.7	M	2.4
T	9.1	W	2.4
A	8.2	F	2.2
O	7.5	G	2.0
I	7.0	Y	2.0
N	6.7	P	1.9
S	6.3	B	1.5
H	6.1	V	1.0
R	6.0	K	0.8
D	4.3	J	0.2
L	4.0	X	0.2
U	2.8	Q	0.1
C	2.8	Z	0.1

usage. The above frequency patterns in Table 3.1 have been noted across large samples of English texts (Elwes 2014: 345).

Frequency analysis is based in large part on N-gram theory, which has become a major theoretical paradigm within cryptanalysis generally. An N-gram model predicts the next item in a sequence of an (n−1)-order Markov chain. It is used commonly in text-messaging devices, which are programmed to predict a complete word as one types it up. The basic idea in this model goes back to the founder of information theory, the American engineer Claude Shannon (1948), who asked the question: "Given a sequence of letters (for example, the sequence '*for ex*'), what is the likelihood of the next letter?" A probability distribution to answer this question can be easily derived given a frequency history of size n: $a = 0.4$, $b = 0.00001$, . . ., where the probabilities of all the "next letters" sum to 1.0.

Actually, this kind of analysis can be traced back considerably in time. It was employed in the ninth century by the Arab scholar, Al-Khindi, as he studied ancient sacred texts, realizing that some letters were used in a specific language more often than others, and that some appeared rarely or not as frequently, and more importantly that this was consistent across texts. Another use of frequency analysis goes back to Mary Queen of Scots. As she was being held in 1586, she developed a cryptographic system to send her letters to a Catholic sympathizer based on twenty-three letter substitutions and thirty-six code symbols for words and phrases. But her correspondent, Gilbert Clifford, was a double agent and he routinely opened her letters in order to give them to English spymaster Francis Walsingham, who set up a cipher school in London in order to break Mary's code. The method developed by the school was based on a simple analysis of the frequency of the verbal structures used by Mary.

The first scientific frequency analysis in cryptography was conducted by Leon Battista Alberti in the 1461 (Alberti 1997) who was the inventor of polyalphabetic ciphers—Caesar ciphers that use more than one alphabet. Compared to other codes used in his era, Alberti's method was unbreakable without knowing the key or code. Largely on the basis of Alberti's work, Johannes Thrithemius published the first systematic book on cryptanalysis titled, *Steganographia,* in 1516. Then, in 1586, Blaise de Vigenère published his famous Vigenère Square—a method of encryption extending Alberti's polyalphabetic code. A table is used with the alphabet written twenty-six times in different rows, with each alphabet shifted cyclically to the left corresponding to twenty-six Caesar ciphers. Encryption is based on using one of the different alphabets, depending on a repeating keyword. The Vigenère Square has been a significant one in the history of cryptography, capturing the imagination of many writers and puzzlists. Lewis Carroll, for instance, considered messages encoded with it to be unbreakable in an 1868 essay. But one of the founders of modern-day computer science, Charles Babbage, broke a variant of the code in 1854, although he apparently did not publish his method (Singh 1999: 63–78). Shortly thereafter, in 1863, Friedrich Kasiski published a general method for solving the Vigenère cipher (Franksen 1985).

The gist of the enigmatological story of cryptograms is that they tap into our sense of mystery, like many other puzzle constructs. They are thus more than games, although they share features with them (as discussed above). Their mystery value is the reason why they have been incorporated into many stories. Cryptogram puzzles play latently on our intrinsic sense of mystery. Concealed messages depend on being unsuspected; once they are discovered, they offer little difficulty towards their decipherment. But until then they leave us in suspense and desiring to unravel them.

Word squares, word searches, crosswords, and doublets

The Sator Acrostic discussed above was a sacred word square. The first use of a word square as a recreational game can likely be traced to the American magazine *Godey's Lady's Book and Magazine* in 1862, although there is some question as to whether the game published there may have been modeled on something originally published in England in 1859. Unlike the Sator Acrostic, the word square puzzle is not palindromic: it contains the words *circle*, *Icarus*, *rarest*, *create*, *lustre*, and *esteem*, as shown in Figure 3.4.

C	I	R	C	L	E
I	C	A	R	U	S
R	A	R	E	S	T
C	R	E	A	T	E
L	U	S	T	R	E
E	S	T	E	E	M

Figure 3.4 The original word square puzzle.

A likely derivative of this genre of puzzle is known as a *word search* puzzle, which simply presents a square array of letters in which actual words are hidden:

The names of five colors are concealed in Figure 3.5. They can be read from left to right, from right to left, upward and downward, and diagonally upward and downward.

A	B	Y	E	L	L	O	W	N	M
H	J	K	L	M	N	O	P	N	B
V	M	N	N	E	E	R	G	G	L
B	N	M	N	K	O	P	L	K	U
S	W	E	I	B	N	M	K	Y	E
R	T	K	P	W	T	Y	U	M	N
Y	O	P	J	N	H	M	L	M	N
A	D	E	R	Q	U	I	I	E	O
N	M	E	L	P	O	M	T	B	V
X	X	X	R	A	S	D	C	E	X

Figure 3.5 A word search puzzle.

The words for the colors *yellow*, *green*, *blue*, *white*, and *red* are inserted in the square as shown in Figure 3.6.

A	B	Y	E	L	L	O	W	N	M
H	J	K	L	M	N	O	P	N	B
V	M	N	N	E	E	R	G	G	L
B	N	M	N	K	O	P	L	K	U
S	W	E	I	B	N	M	K	Y	E
R	T	K	P	W	T	Y	U	M	N
Y	O	P	J	N	H	M	L	M	N
A	D	E	R	Q	U	I	I	E	O
N	M	E	L	P	O	M	T	B	V
X	X	X	R	A	S	D	C	E	X

Figure 3.6 Solution to the word search puzzle.

Word search puzzles are very popular today and are also commonly used as pedagogical tools; that is, as games for teaching children to identify the alphabetic structure of words. The first such game appeared in the March 1, 1968 issue of the *Selenby Digest* in Norman, Oklahoma, catching on quickly, as teachers in the local schools started asking for reprints of the puzzle for their classes. As Hovanec (1978: 24) observes, games of this kind emerged after the advent of print technology in the mid-1400s, becoming part of recreational culture broadly:

> Puzzles began to crop up in books and magazines, and their popularity grew at an amazing rate. And since the advent of mass publishing, a voluminous number of puzzles have been retained. A study of these puzzles shows a remarkable parallel to the growth of literature. Puzzles in the sixteenth through nineteenth centuries were considered artistic manifestations of the creative mind and were sometimes seen to rival the best literary works of the times.... An entirely different outlook exists today. Puzzle solvers view their hobby as an amusing and somewhat educational way to pass the time.

In the Internet Age, word games have migrated to online culture where they are among the most accessed recreational sites in the virtual world. Again, this corroborates the existence of an implicit principle in the flow of enigmatological history—namely, a shift from mythic to ludic and literary. Indeed, there are now countless sites for puzzles such as anagrams, which automatically convert words and phrases into other words and phrases. Here they are perceived purely as recreational devices, not as having prophetic value, as they once did.

Of all the word games discussed so far in this chapter, the *crossword* is likely to be the most popular of all today. It is a twentieth-century invention, devised not for some mystical or divinatory reason, but for the sole purpose of providing intellectual entertainment. It was created by Arthur Wynne, who was born in Liverpool, England, in 1871, immigrating to the United States in 1905. As editor of the "Fun Page" of the *New York World*, Wynne introduced what he had intended to call a *word cross* on December 21, 1913, after he had seen something similar to it in England. Because of an error in typesetting, the puzzle appeared with the title *crossword* instead. That name has stuck ever since. Figure 3.7 shows Wynne's original crossword (the word "fun" was given as a "free word"). The words are to be guessed from the given clues and inserted correctly into the interlocking grid of horizontal and vertical squares.

As can be seen, crossword puzzles involve an interplay between word and phrase meanings and their alphabetic structure. Unexpectedly, readers inundated

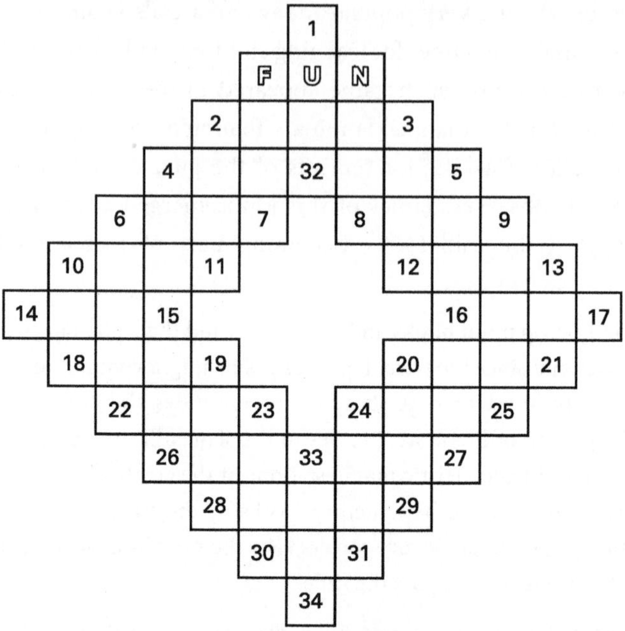

Fill in the small squares with words which agree with the following definitions:

2-3. What bargain hunters enjoy.
4-5. A written acknowledgement.
6-7. Such and nothing more.
10-11. A bird.
14-15. Opposed to less.
18-19. What this puzzle is.
22-23. An animal of prey.
26-27. The close of a day.
28-29. To elude.
30-31. The plural of is.
8-9. To cultivate.
12-13. A bar of wood or iron.
16-17. What artists learn to do.
20-21. Fastened
24-25. Found on the seashore.

10-18. The fibre of the gomuti palm.
6-22. What we all should be.
4-26. A day dream.
2-11. A talon.
19-28. A pigeon.
F-7. Part of your head.
22-30. A river in Russia.
1-32. To govern.
33-34. An aromatic plant.
N-8. A fist.
24-31. To agree with.
3-12. Part of a ship.
20-29. One.
5-27. Exchanging.
9-25. Sunk in mud.
13-21. A boy.

Figure 3.7 The original crossword puzzle (from Wikimedia Commons).

Wynne was inundated with requests for more puzzles. Overnight, the crossword had become a puzzle craze in the city of New York. By 1924, crosswords had grown into a national pastime in America, with the grid gradually taking on the more familiar appearance of patterned black and white cells, with two sets of clues, one for words to be written horizontally *(across clues)* and the other for words to be written vertically (*down clues*). In 1924, the American publishing company Simon and Schuster printed the first book compilations of crossword puzzles. Each book came equipped with a pencil, eraser, and a penny postcard, which buyers could mail to the publisher to request the answers. The first book alone sold nearly half a million copies. To take advantage of the crossword mania, manufacturers soon began making jewelry, dresses, ties, and other objects with crossword designs on them. A song called "Crossword Mama, You Puzzle Me, but Papa's Gonna Figure You Out," hit the top of the charts in 1924. A Broadway play, *Puzzles of 1925*, which dealt with the crossword puzzle craze, met with resounding success. The heart of the play featured a scene in a "Crossword Puzzlers Sanitarium" for people driven insane by their obsession with crossword puzzles. Shortly thereafter, board games were invented incorporating the crossword puzzle idea. Of these, Scrabble has remained the most popular.

The crossword was really no more than a descendant of the acrostic puzzle concept, as Wynne himself acknowledged (quoted in Costello 1988: 22):

> I awakened recently to find myself acclaimed as the originator of the crossword puzzle, which everybody is doing now. But all I did was take an old idea as old as language and modernize it by the introduction of black squares. I'm glad to have had a hand in it, and no one is more surprised at its amazing popularity.

Today, versions and varieties of the crossword are numerous. But all are based on the acrostic principle described in this chapter—words crossing each other in specific ways. The innovative feature of Wynne's puzzle was the inclusion of clues and the fact that each letter in a word has to be crossed by one other. In early crosswords, clues consisted of simple synonyms, antonyms, or primary definitions. The level of difficulty rose quickly, however, as skilled crossword solvers started demanding much more challenging clues. Special dictionaries soon had to be compiled of unusual words that had found their way into crossword clues.

The three individuals who developed the crossword into its contemporary forms were Margaret Petherbridge Farrar, Edward Powys Mather, and Elizabeth S. Kingsley. Farrar became puzzle editor of the *New York World* in

1921, producing ingenious, challenging crosswords that garnered a large readership. She was also the author of the first few Simon and Schuster crossword puzzle books. By the fall of 1924, these were among the five best-selling nonfiction books. The royalties, incidentally, allowed her to loan the editor John Farrar, whom she had married, sufficient capital to start two publishing houses, Farrar and Rinehart in 1929 and Farrar, Straus and Company in 1946 (which later became Farrar, Straus & Giroux). She became the first crossword puzzle editor at *The New York Times* in 1942, holding that post until 1969; in the same period, she also created crosswords for such popular magazines as *Esquire*, *Seventeen*, *Sports Illustrated*, and *Good Housekeeping*. She was responsible for standardizing the symmetrical grid format that most crossword puzzles have today, with black and white boxes and across and down clues. She also introduced the now common practice of including quotations, puns, and all kinds of word play in the clues.

The British crossword-maker Edward Powys Mather, known to his puzzle audience as Torquemada (the name of a notorious Spanish Grand Inquisitor), introduced a very difficult version, which consists of clues that are themselves puzzles. For instance, a clue might hide the required word in its definition; so, if the word is *nylon*, the clue might be, *Material used in many long dresses*. This type of clue within the clue made its way into American crosswords after Broadway composer Stephen Sondheim launched a brief career as a puzzle writer in 1968, creating Torquemada-like crosswords for the *New York* magazine. Shortly thereafter, the "cryptic crossword," as it came to be known, became one of the most popular of all among crossword aficionados. To grasp the ingenious nature of cryptic crosswords, consider the following clues:

Used to be healthy, so breathe out: "used to be" = ex and "healthy" = hale (answer: exhale)

Elvis in blue jeans (answer: Levis, which are "blue jeans" and an anagram of Elvis)

The artist has been about cooked with herbs
(answer: saged)

Marriage days
(answer: weds)

Tell the physician to love, and then he's quite often my subject
(answer: drama)

We can't honestly say there's no harm in it, but we're glad the French girl's got it
(answer: charm)

A cryptic clue explains the answer, while at the same time providing a confusing surface reading. All kinds of word play are involved in this type of crossword, from puns, anagrams, and homophones, to idiomatic plays on phrases and double entendre. For example, the answer to the clue *Cat's tongue* is *Persian*, which is a type of cat as well as a tongue (the Persian language). Mather composed crossword puzzles for the Saturday *Westminster Gazette* in 1925 and then for the *Observer* from 1926 to 1939. He soon gained a reputation as one of the toughest crossword puzzlists of his era.

Elizabeth S. Kingsley introduced the *double-crostic* in 1934 in the *Saturday Review of Literature*. The initial idea can be traced to a poem that Lewis Carroll wrote to Mabel and Emily Kerr on May 20, 1871 (Wakeling 1992). He titled it "A Double Acrostic:"

Thanks, thanks, fair Cousins, for your gift
So swiftly borne to Albion's isle
Though angry waves their crests uplift
Between our shores, for many a league!

("So far, so good," you say: "but how
Your Cousins?" Let me tell you, Madam.
We're both descended, you'll allow,
From one great-great-great-grandsire, Noah.)

Your picture shall adorn the book
That's bound, so neatly and moroccoly,
With that bright green which every cook
Delights to see in beds of cauliflower.

The carte is very good, but pray
Send me the larger one as well!
"A cool request!" I hear you say.
"Give him an inch, he takes an acre!"

"But we'll be generous because
We well remember, in the story,
How good and gentle Alice was,
The day she argued with the Parrot!"

The last word in each second line of the five stanzas is a ruse. If we replace that word with a word that fits into the rhyme pattern, and then list the five words separately, we get the following acrostic layout in which the first and last letters spell MABEL and EMILY:

Mile (rhymes with "isle")
Adam (rhymes with "Madam")
Broccoli (rhymes with "moroccoly")
Ell (rhymes with "well")
Lory (rhymes with "story")
First letters: M + A + B + E + L = Mabel
Final letters: E + M + I + L + Y = Emily

Kingsley's version involves guessing words defined in the puzzle and then writing them in numbered squares in a diagram, in which a quotation, the name of a famous personage, the title of some work, and the like comes into view when the correct letters are supplied. Today, the most acclaimed crosswords are those found in *The New York Times*. Since 1993, they have been edited by Will Shortz.

The crossword puzzle has unconscious appeal, perhaps because it creates a reassuring sense of order as we fill the void (the empty cells) with words. The grid, with its symmetrical rectangular configuration crying out for missing letters, has hooked millions across the globe, who might see the crossword as a means to fill in the emptiness. Many stories are told, in fact, of crossword puzzle addiction, arguably for this reason. According to one story, two New York magistrates had to ration crossword addicts who had neglected their families to two puzzles a day (Costello 1988: 23). Another tells of a human life put in jeopardy because of crossword-puzzle deprivation (Hovanec 1978: 109):

> During one especially debilitating snowstorm when service to rural areas was completely cut off, a nurse at a hospital in one of the affected areas wrote to Mrs. Farrar and told her that a patient's recovery was in jeopardy unless she received a copy of the answers to a previous puzzle. It seemed that this patient was recuperating through solving the crossword found in the *Times*.

Word game crazes like the crossword are not the prerogative of the twentieth century. As discussed, anagrams became a sort of craze as a literary genre (above). But perhaps the first true word game craze can be traced to Lewis Carroll's *doublet* puzzle. This consists of two words whereby we must transform one into the other by changing only one letter at a time, forming a legitimate new word with each change. Carroll published many doublet puzzles in *Vanity Fair* between 1879 and 1881 (Wakeling 1992: 39–41). He produced his first doublet for the March 29, 1879 issue of the magazine:

> *Turn the word HEAD into TAIL, by changing only one letter at a time, forming a new word each time to do so.*

Here is the solution:

Head
↓
Heal
↓
Teal
↓
Tell
↓
Tall
↓
Tail

The puzzle became so popular among readers that the magazine offered weekly prizes for solutions. Carroll later amended the rules of his game slightly in order to add variety and, thus, stimulate more interest, as can be seen in the following puzzle he created in 1892:

Change IRON into LEAD by introducing a new letter or by rearranging the letters of the word, at any step, instead of introducing a new letter. But you may not do both in the same step.

Here's the solution:

Iron
↓
Icon (new letter)
↓
Coin (rearrangement)
↓
Corn (new letter)
↓
Cord (new letter)
↓
Lord (new letter)
↓
Load (new letter)
↓
Lead (new letter)

Carroll took the name for his puzzle from the witches' incantation in *Macbeth*: "Double, double, toil and trouble." He had initially named it, "Word-links," in his pamphlet, *Word-links: A Game for Two Players, or a Round Game*, published in April of 1878. In it, he states that he had invented the game for two girls to play on Christmas Day of 1877, because they usually "found nothing to do." Given the popularity of the doublets, in 1879 the publisher Macmillan put out a thirty-nine-page booklet containing several of the puzzles, titled *Doublets: A Word Puzzle*. This became highly popular and was increased to seventy-three pages in a second edition, published in 1880. A third edition, put out in the same year, was enlarged even more to eighty-five pages.

In his classic word game collection of 1965, *Language on Vacation*, Dimitri Borgmann renamed the doublet, *Word Ladder*, which concretely describes the solution strategy itself, consisting of "word steps" that are connected logically as are the steps in a real ladder.

The ludic nature of language

Word games are experiments in the relation between language and ludic cognition. Lewis Carroll was fascinated by the boundless range of meanings that language has the potential to encode. By showing how one small change in a word leads to a completely new set of meanings in his doublets concept, he displayed how language can do so much with a limited set of resources—around fifty phonemes represented by twenty-six alphabet characters. As two twentieth-century linguists, Edward Sapir (1921) and Benjamin Lee Whorf (1956), showed in their groundbreaking research, human ideas are, as Carroll suspected, rooted in the structure of language. When we create a word, we end up believing that the things it refers to must exist, in reality or in the imagination. In actual fact, words are selections of all that can be experienced and perceived. By playing with them, we become consciously aware of the tenuous connection between words and what is real.

In no way does this imply that language constrains or stifles the imagination. On the contrary, the riddles, anagrams, acrostics, cryptograms, and other linguistic puzzle artifacts that have marked human history make it obvious that language is a malleable instrument that can be put to any use the human mind desires. Should the need arise to create a new word-category, all we have to do, as Carroll showed in his poem *Jabberwocky*, is to be consistent with the structural requirements of our language's sound system, changing one phoneme at a time.

Fascinated by the verbal side of *homo ludens*, an association called Oulipo—an acronym for "*Ouvroir de littérature potentielle*" ("Workshop of Potential

Literature")—was founded in Paris in the mid-twentieth century by a small group of writers and mathematicians devoted to exploring the relation between language and reality (Paulos 1991: 166–8). Their activities were (and continue to be) truly intriguing, and highly relevant to the subject matter of this chapter. For example, Raymond Queneau, one of the founders, published a book of poetry titled *100 Trillion Sonnets*, consisting of ten sonnets, one on each of ten pages. The pages are cut so as to allow each of the fourteen lines of a sonnet to be turned separately. The physical format of the book allows 100 trillion combinations of lines—100 trillion sonnets. All of them, Queneau claimed, "make sense." Another example of an Oulipian work is Georges Perec's 300-page novel *La disparition*. None of the words in the novel contain the letter *E*. This type of work is called a *lipogram*. Perec insisted that his lipogrammatic novel was worthy of being considered literature, since it was designed to explore language's infinite possibilities. Incidentally, the celebrated American humorist James Thurber also wrote a lipogrammatic work in 1957, titled *The Wonderful O*. It is a political fable for children, telling what happens when Captain Black, a literate pirate who hated the letter *O*, banished the letter from the island of Ooroo.

Another Oulipian technique is to make up sentences, called *pangrams*, which contain all the letters of the alphabet. Here are a few examples of pangrams (Bombaugh 1961: 326):

Pack my box with five dozen liquor jugs (32 letters)
A quick brown fox jumps over the lazy dog (33 letters)
Waltz, nymph, for quick jugs vex Bud (28 letters)
Quick wafting zephyrs vex bold Jim (29 letters)

Lipograms and pangrams are ludic artifacts showing that a constraint on linguistic structure in no way limits the ability of words to create or encode meanings. The ancient anagrammatists sought to uncover hidden meanings in virtually the same way—by playing with the rearrangement of letters. Interestingly, around 2,500 years ago, the Greek poet Pindar (*c.* 522–443 BCE) created puzzle poems that followed a certain rule or bore a hidden message (Costello 1988: 5). For example, he wrote an ode without using the letter *sigma*. He was thus the first Oulipian, long before the twentieth century. As Wittgenstein (1953) came to understand, we play the game of language every day. We are impelled to do so by *homo ludens* within us.

4
Visual Puzzles

Life proceeds as a series of optical illusions, artificial needs and imaginary sensations.

Alexander Herzen (1812–70)

The English expression, "Seeing is believing," constitutes an idiomatic portrayal of a psychological theory of perception. Consider the classic optical illusion, called an ambiguous figure (Figure 4.1)

How many figures do you see in the image below?

Figure 4.1 The Rubin illusion (from Wikimedia Commons).

The illusion is based on a figure-ground confusion. Are the dark faces that one sees at certain instances of concentration part of the white goblet that comes into focus at other instances, or part of the background? Our brain seems to interpret the visual information in the drawing in the same manner that we process semantic ambiguities in language, creating a double-vision effect that is analogous to the double-entendre effect (Luckiesh 1965). The picture was devised by the Dutch psychologist Edgar Rubin in 1915. He explained the illusion as follows:

> One can then state as a fundamental principle: When two fields have a common border, and one is seen as figure and the other as ground, the immediate perceptual experience is characterized by a shaping effect which emerges from the common border of the fields and which operates only on one field or operates more strongly on one than on the other.

This type of optical illusion has intrigued psychologists, artists, mathematicians, and others. The great painter Maurits C. Escher utilized similar figure-ground ambiguous effects in several of his works to considerable aesthetic effect. An example is his woodcut *Day and Night* (1938). In the picture, the birds seem to come out of the landscape at one viewing and, vice versa, the landscape emerges at another viewing, as the birds seem to mesh with it.

Are such optical illusions puzzles, as defined in this book? They certainly seem to play on visual perception in an analogous way in which word games play on verbal perception. The Q (Question) in this case is: *What do you see*? The A (Answer) is not obvious and when it does come, literally, into view, it produces a similar kind of Aha Effect of all puzzles. For this reason, they can indeed be classified as puzzles that play on the relation between vision and perception. This chapter deals with all kinds puzzles that involve visual perception and visual logic, selectively, from the ancient geometrical conundrums and games, to rebuses and mazes. Visual puzzles bear implications for the nature of human perception and its relation to cognition, as well as for the origin of consciousness itself.

Optical illusions

Optical illusions hold a warning about visual perception—things are not always what they appear to be. They do so by tricking the eye into seeing images incorrectly, inducing us to interpret them ambiguously, or else into perceiving a figure as representing something that is physically impossible. Figure 4.2 is the classic example of an image that leads to an incorrect interpretation:

Which line is longer?

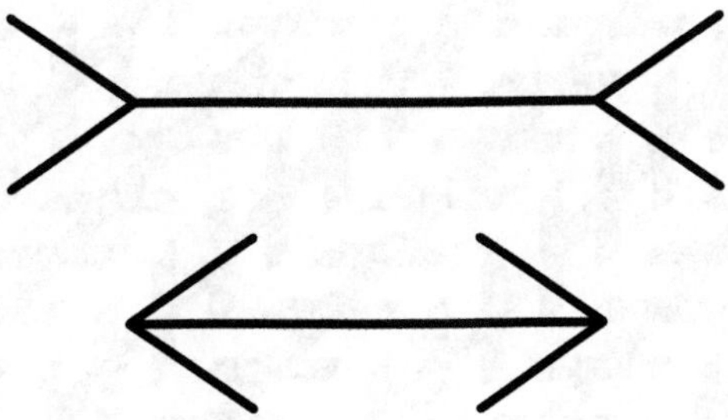

Figure 4.2 The Müller-Lyer illusion.

We tend to see the top line segment as longer than the one below it, even though the lines are equal in length. People reared in different cultures are not so easily fooled by the illusion, called the Müller-Lyer Illusion, after the German sociologist Franz Carl Müller-Lyer devised it in 1889. The source of the illusion is the different orientations of the two arrowheads, which dupe the eye into perceiving one line as longer than the other. They are visual counterparts to the conceptual metaphor: *outward expansion means growth, inner expansion means shrinkage.* In effect, the illusion is connected to metaphorical cognition and its cultural embedment, as the relevant research appears to suggest. For instance, the illusion does not work in cultures where the architecture is not rectilinear but of some other shape (Berry 1968). Research has also shown that it does not work in cultures where Euclidean geometry has historically played little or no role, such as in the culture of the San foragers of the Kalahari Dessert, who do not assign the same interpretation to the figures as we do. Explanations other than the cultural one have come forward, such as the view that eye pigmentation has a role to play in the illusion, but these have not held up to further research scrutiny and, thus, need not concern us here.

A similar deceptive effect on perception comes from the Zöllner Illusion, discovered by astrophysicist Johann Zöllner, who, in 1860, stumbled upon a piece of fabric with a design that made parallel lines appear decidedly unparallel to the eye. Each of the lines in Figure 4.3 are parallel to each other, but they do not seem to be so. They are an example of the Zöllner Illusion.

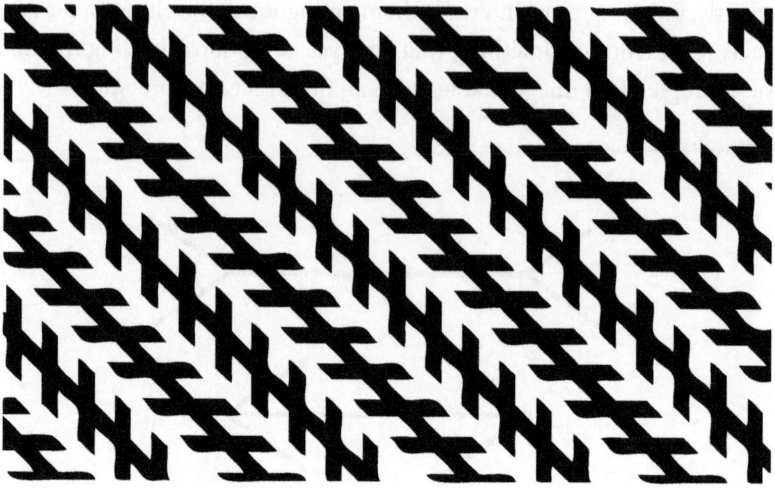

Figure 4.3 The Zöllner illusion (from Wikimedia Commons).

The source of the illusion is the addition of smaller slanted lines on the parallel lines, which we interpret as slanted because we are conditioned again by a metaphorical suggestion—*slanting lines entail distortion*. The fact that the brain is duped with such optical illusions to see what is not actually there suggests, paradoxically, that the brain is not a pre-programmed information-processing device; rather, it is a plastic organ that is influenced by external stimuli and factors. The above illusions also suggest that the same neural structures involved in producing metaphor are involved in shaping perception.

The plasticity of perception is even more conspicuous in the case of so-called *impossible figures*. Consider the staircase in Figure 4.4, which seems to defy the laws of physics as it is drawn, since the stairs seem to go both up and down—if we start climbing at one end of the staircase, by the end we will end up back at the starting point; vice versa, starting at one end and descending the staircase, we will also end up at our starting point. Although it was discovered by Swedish artist Oscar Reutersvärd in 1915, the illusion was made famous by Lionel S. Penrose and his son Roger Penrose (1958).

Illusions of this type are the products of linear perspective drawing that reaches back to the ideas and techniques of Renaissance architect and artist Filippo Brunelleschi, who developed a mathematical system for showing depth on a flat surface, based on the technique that parallel lines seem to converge as they recede toward a vanishing point. Artists in antiquity also knew the principles of linear perspective, but their ideas were largely abandoned during the Middle Ages,

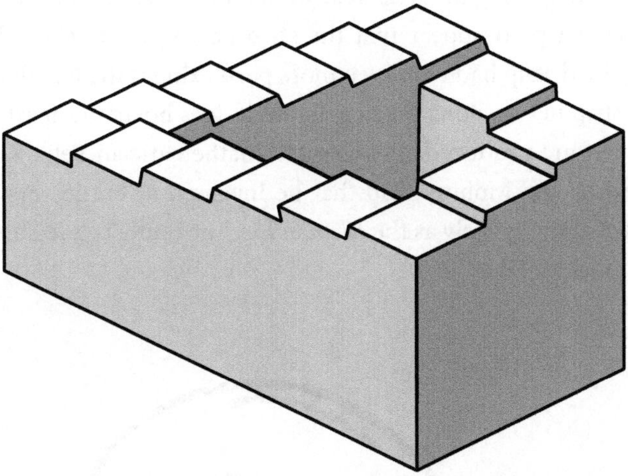

Figure 4.4 Impossible staircase by Penrose and Penrose (1958) (from Wikimedia Commons).

being rediscovered during the Renaissance, after Leon Battista Alberti wrote *On Painting* (1435), the first scientific study of linear perspective.

Optical illusions show, as Shepard (1990: 168) aptly puts it, that our "visual experience evidently is the product of highly sophisticated and deeply entrenched inferential principles that operate at a level of our visual system that is quite inaccessible to conscious introspection or voluntary control." One of the most famous of all illusions, actually, comes out of mathematics. It is the *Möbius strip*, invented by the German astronomer and mathematician, August Ferdinand Möbius in 1858 (see Pickover 2005). To make such a figure, we take a strip of paper, give it a half twist, and join the ends (Figure 4.5).

Figure 4.5 Möbius strip (from Wikimedia Commons).

Tracing a pencil along the strip at any point brings us right back to where we started. Thus, it would appear that the strip has only one side, although the original unjoined strip had two. Even more perplexing is the fact that if one cuts the Möbius strip in two along the pencil line, it does not come apart—two strips linked together are produced. The German mathematician Felix Klein became so captivated by the Möbius strip that he invented a "bottle version" of it in 1882—known appropriately as the *Klein bottle*. The bottle is one-sided. It has no ends and no inside (Figure 4.6).

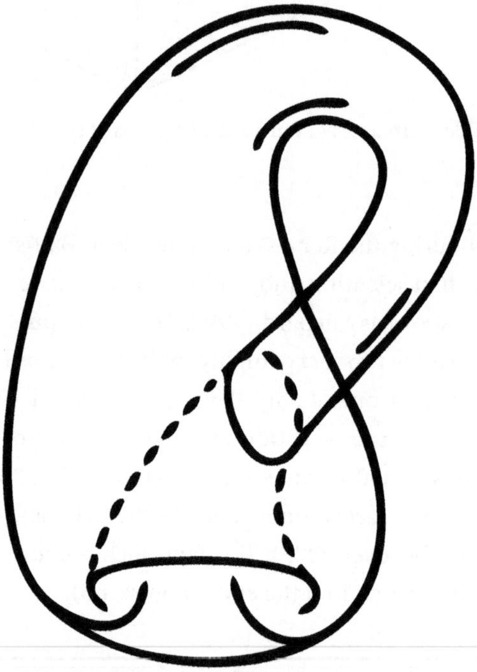

Figure 4.6 Klein bottle (from Wikimedia Commons).

The most prominent artist in drawing all kinds of optical illusions is, as mentioned, Maurits Escher. Beginning in about 1936, Escher started drawing ambiguous objects in which he interlocked repeated figures of stylized animals, birds, or fish, leaving no spaces between the figures (Schattschneider 1990). A little later, he began experimenting with the brain's plastic perception system, creating such illusions as staircases that appeared to lead both upward and downward in the same direction (like the illusion in Figure 4.4), and alligators that seemed to come to life, walking off the edge of the paper.

Oddities such as the Möbius strip and the Klein bottle are studied under the rubric of *topology*, which concerns itself with determining the *insideness* or *outsideness* of shapes. A circle, for instance, divides a flat plane into two regions, an inside and an outside. A point outside the circle cannot be connected to a point inside it by a continuous path in the plane without crossing the circle's circumference. If the plane is deformed, it may no longer be flat or smooth, and the circle may become a crinkly curve, but it will continue to divide the surface into an inside and an outside. A knot, too, may be viewed as a simple closed curve that can be twisted, stretched, or otherwise deformed, but not torn (Neuwirth 1979). Two knots are equivalent if one can be deformed into the other; otherwise, they are distinct.

Perspective drawing and the oddities of perception it produces has actually been useful for reasoning about n-dimensional space, as the preacher and literary critic Edwin A. Abbott showed in his remarkable novel of 1884, *Flatland: A Romance of Many Dimensions*. The characters of the novel are geometrical figures living in a two-dimensional universe called Flatland. They see each other edge-on, and thus as dots or lines, even though, from the vantage point of an observer in three-dimensional space looking down upon them, they are actually lines, circles, squares, triangles, and so on. To grasp the difference that the viewing perspective makes, Flatland can be imagined as the surface of a table. If one crouches to look at a square piece of paper lying on the table, with eyes level with the surface, it will appear as a line (or a point if our line of sight is on one of its vertices). The only way to see it as a square is to view it from above the table.

In his first chapter of Part I ("This World"), Abbott (2002: 34–5) provides a description and an accompanying graphic to describe Flatland. It is worthwhile to reproduce it here since it encapsulates what his objective was:

> I call our world Flatland, not because we call it so, but to make its nature clearer to you, my happy readers, who are privileged to live in Space.
>
> Imagine a vast sheet of paper on which straight Lines, Triangles, Squares, Pentagons, Hexagons, and other figures, instead of remaining fixed in their places, move freely about, on or in the surface, but without the power of rising above or sinking below it, very much like shadows—only hard and with luminous edges—and you will then have a pretty correct notion of my country and countrymen. Alas, a few years ago, I should have said "my universe": but now my mind has been opened to higher views of things.
>
> In such a country, you will perceive at once that it is impossible that there should be anything of what you call a "solid" kind; but I dare say you will suppose

that we could at least distinguish by sight the Triangles, Squares, and other figures, moving about as I have described them. On the contrary, we could see nothing of the kind, not at least so as to distinguish one figure from another. Nothing was visible, nor could be visible, to us, except Straight Lines; and the necessity of this I will speedily demonstrate.

Place a penny on the middle of one of your tables in Space; and leaning over it, look down upon it. It will appear as a circle.

But now, drawing back to the edge of the table, gradually lower your eye (thus bringing yourself more and more into the condition of the inhabitants of Flatland), and you will find the penny becoming more and more oval to your view; and at last when you have placed your eye exactly on the edge of the table (so that you are, as it were, actually a Flatlander) the penny will then have ceased to appear oval at all, and will have become, so far as you can see, a straight line.

The same thing would happen if you were to treat in the same way a Triangle, or Square, or any other figure cut out of pasteboard. As soon as you look at it with your eye on the edge on the table, you will find that it ceases to appear to you a figure, and that it becomes in appearance a straight line. Take for example an equilateral Triangle—who represents with us a Tradesman of the respectable class. Fig. 1 represents the Tradesman as you would see him while you were bending over him from above; figs. 2 and 3 represent the Tradesman, as you would see him if your eye were close to the level, or all but on the level of the table; and if your eye were quite on the level of the table (and that is how we see him in Flatland) you would see nothing but a straight line.

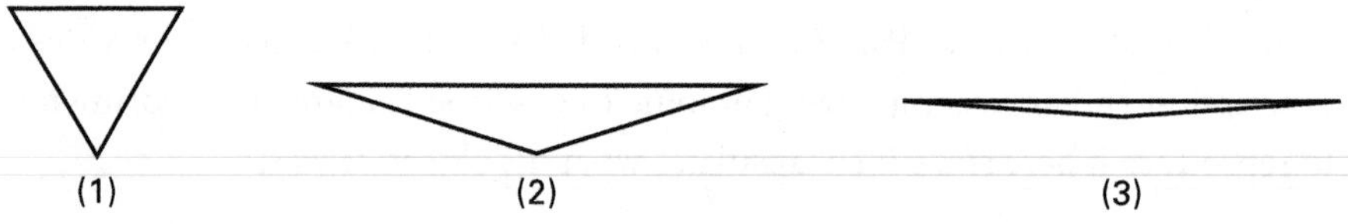

This type of perspective thinking has allowed geometers to entertain truly intriguing questions about space: Is there a formal relation between our three-dimensional world—Sphereland—and the two-dimensional one? The answer is yes, because a Sphereland figure can be changed into a Flatland one, and vice versa, by performing a specific kind of alteration to it. Consider a three-dimensional box—a cube made up of six sides, which are, geometrically speaking, six equal Flatland squares. This can be easily transformed into a Flatland figure, as shown in Figure 4.7, by simply "unfolding" the six square sides onto a flat surface.

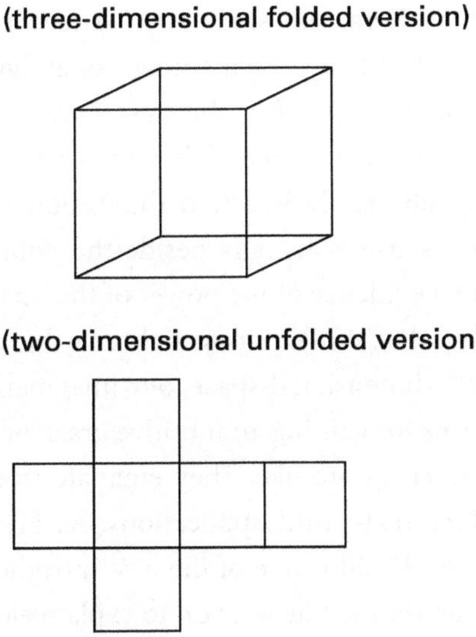

Figure 4.7 Unfolding a cube.

The resulting Flatland figure can, of course, be just as easily transformed back into a three-dimensional cube by folding the six squares together. Now, the question that such folding and unfolding raises is truly mind-boggling: Can an analogous transformation be envisioned that would relate to a four-dimensional cube? Using the same kind of reasoning, geometers and artists have produced a hypercube, which is defined as a four-dimensional cube, as shown in Figure 4.8.

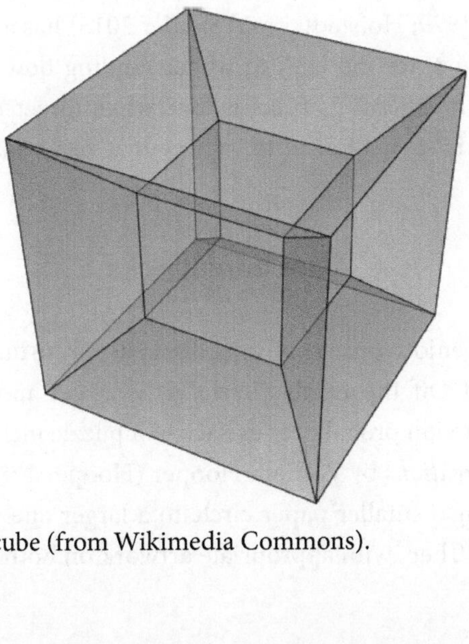

Figure 4.8 A hypercube (from Wikimedia Commons).

It is composed of eight cubes, sixteen vertices, twenty-four squares, and thirty-two edges. But is it truly a four-dimensional figure? Like the Flatlanders, we Spherelanders can only infer, through our imagination, what a four-dimensional world might look like. We will never be able to truly *see* one. Whether the hypercube really is a four-dimensional structure or not in any meaningful physical sense is actually beside the point. The fact that it can be envisioned is indirect evidence of the power of the imagination to think outside the physical box of reality. While our eyesight may be limited to seeing things as they appear in three-dimensional space, our imagination is not. Abbott's novel thus provides a means for gaining an intuitive grasp of what n-dimensional and non-Euclidean geometries are like. They emanate from the imagination, even though they produce real-world applications. As Freiberger (2006) has aptly observed, Abbot's novel "still is one of the best introductions to a mathematical world of higher dimensions." She goes on to explain eloquently why this is so:

> This is an extremely well-thought-out story; every aspect of life in Flatland is accounted for, from housing and climate to the way in which Flatlanders recognise each other's shape (being unable to see their world from above). Abbott's descriptions of how the square manages to imagine a three-dimensional world are a great guide to how we might go about imagining four (or more) dimensions. In fact, the square even conjectures that a four-dimensional world might exist, much to the annoyance of the sphere which considers itself supreme. Abbott's analogy is clear and strong, and will make sense to the most unmathematically minded reader. It's a beautiful defense of mathematical thought and its power to open doors to fascinating new worlds.

As Hofstadter (1979; Hofstadter and Sander 2013) has argued cogently, such imaginative analogies are the keys to understanding how mathematical ideas come into being. The amazing aspect is that, when directed back to the world, they produce theories that allow us to explore that very world.

Vanishing tricks

One of the most ingenious puzzles designed specifically to dupe visual perception is Sam Loyd's "Get Off the Earth Puzzle." It is a "cut-and-slide" trick, whose underlying construction probably goes back to a puzzle included in a 1774 book titled *Rational Recreations* by William Hooper (Hooper 1782). Loyd created his version by fastening a smaller paper circle to a larger one with a pin so that it could spin around. Then, with appropriate artwork on both circles, he made the figure look like the Earth, with thirteen "Chinese warriors" on it. Loyd patented his puzzle in 1897. It sold more than 10 million units in that year alone:

> *When the smaller circle is turned slightly, as shown below, the thirteen warriors turn mysteriously into twelve. Where did the thirteenth warrior go?*

With the world oriented so that the large arrow on it points to the N.E. point on the background, thirteen men can be counted. But when the Earth is turned slightly, so that the arrow points to the N.W. point, there are only twelve such men. What happened to the thirteenth man? Figure 4.9 shows Loyd's presentation of the puzzle.

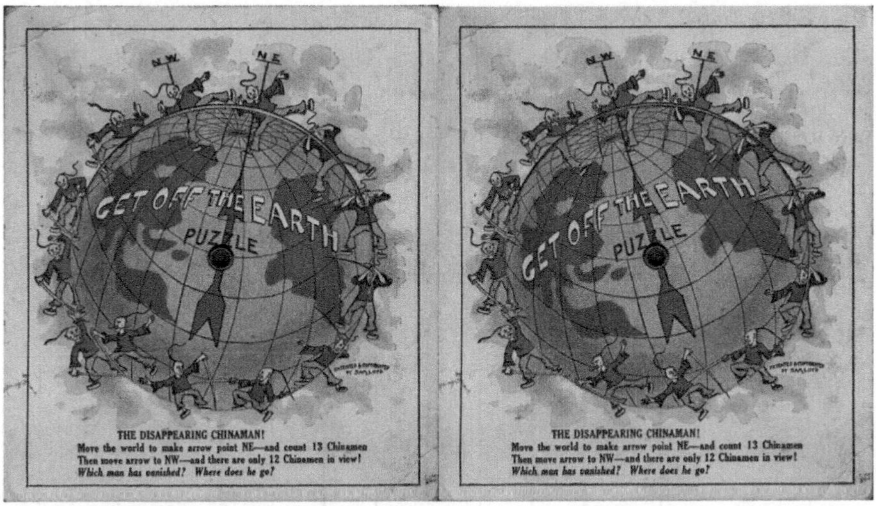

Figure 4.9 Loyd's Get Off the Earth puzzle.

When the Earth is rotated, the pieces are rearranged in such a way that each of the "Chinese Warriors" gains a sliver from his neighbor. For example, at the lower left, there are two Warriors next to each other. The top one is missing a foot. When the Earth is rotated, he gains a foot from his neighbor on the right. That neighbor gains two feet (since he lost one) and one small piece of a leg. And so on and so forth. As a result of the rotation, one of the Warriors will "lose" all his parts, making it seem that he has "disappeared."

To grasp the clever idea underlying the puzzle, it is useful to consider an analogous vanishing trick. Consider a rectangle, ABCD, containing ten straight equidistant parallel lines within it. The rectangle is crossed by a dotted diagonal (Figure 4.10).

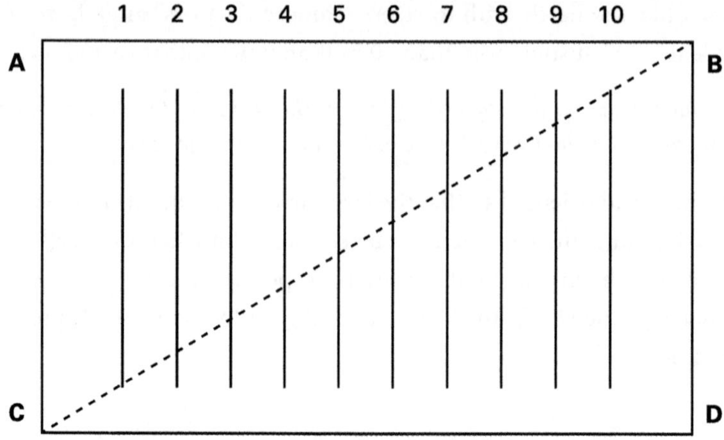

Figure 4.10 Initial diagram of a disappearing line figure.

As can be seen in Figure 4.10, the diagonal touches the top point of line ten and the bottom point of line one. If we cut the rectangle along the dotted line, we will produce an upper and a lower part (Figure 4.11).

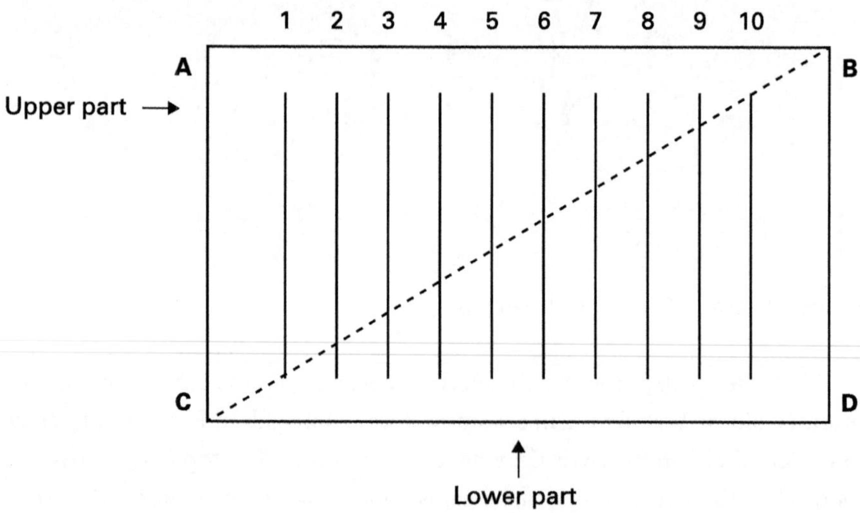

Figure 4.11 Figure shown with upper and lower parts.

Next, we erase the numbers and the letters and we slide the lower piece down and to the left, only to the extent that the linear segments in the lower piece are "in synch" with the linear segments in the upper piece (see Figure 4.12). In this way, we have seemingly preserved the internal lines in the rectangle. Two "protruding lines" are however produced.

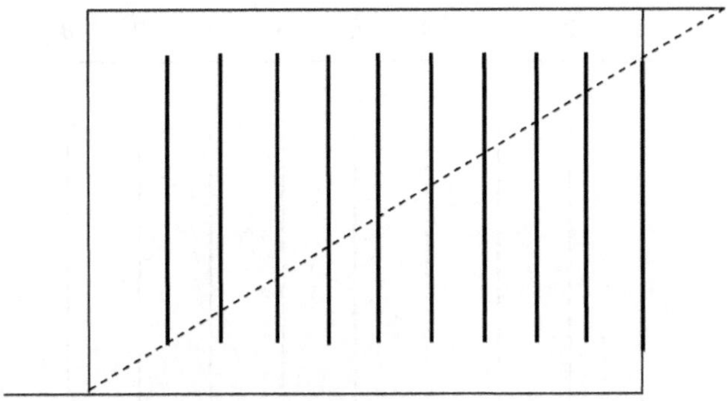

Figure 4.12 Figure resulting from a slide.

Let's cut out the two lines and reassemble the figure anew with a diagonal. This produces a new and slightly smaller rectangular figure (Figure 4.13).

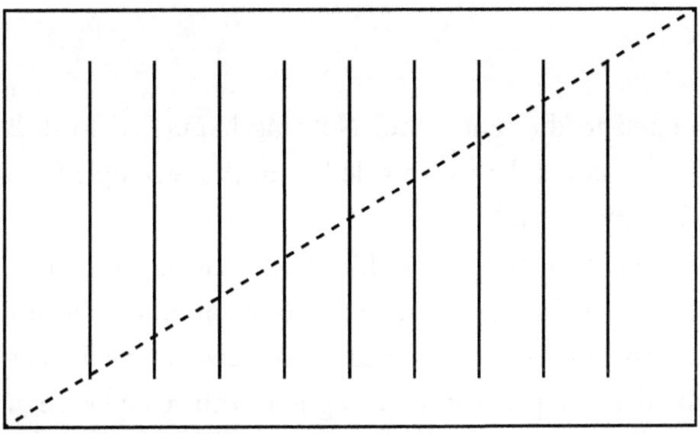

Figure 4.13 Reassembled figure.

If we renumber and re-letter the new figure, we notice that there are now only nine internal lines in the rectangle (Figure 4.14).

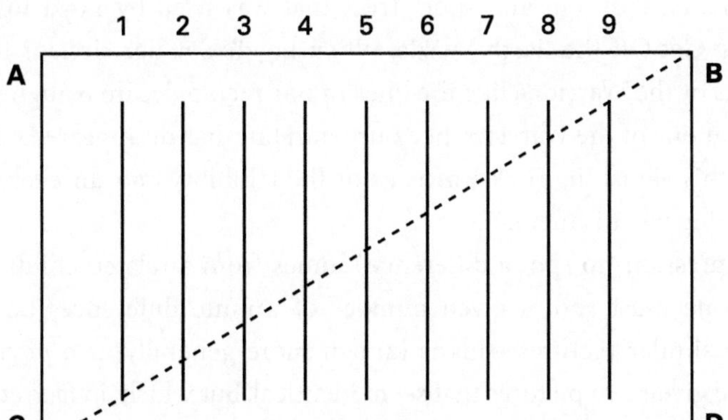

Figure 4.14 Renumbered figure.

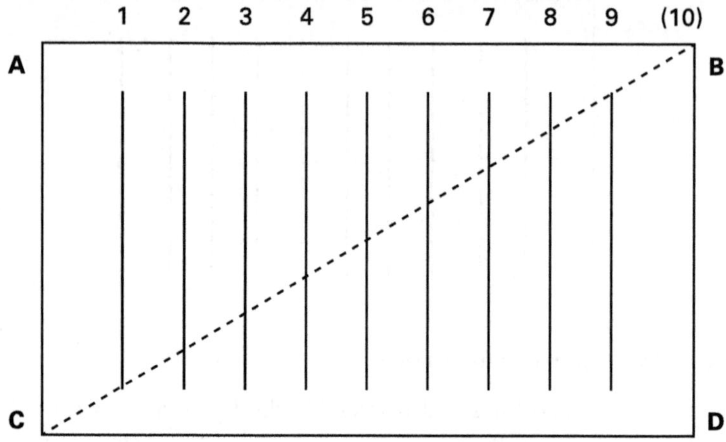

Figure 4.15 Disappearance explained.

What happened to the tenth line? Nothing happened to it. Because of the slide, it has become coincident with side BD of the rectangle (Figure 4.15). It is, in effect, "hidden" by that side.

Let's analyze our vanishing trick. Lines one and ten remain the same after the cut, while the remaining eight lines (two to nine) are sliced into segments. When we slide the lower piece, we produce new internal lines. Each one is now made up of its upper segment aligned with a lower segment that was previously part of the line to the immediate right (as we look at the diagram). The tenth line is still there, but it is now coincident with side BD of the new rectangle. Indeed, if we slide the lower part back up again, the tenth line will reappear.

It is this type of "cut-and-slide" trick that was used by Loyd to create his mystifying Get Off the Earth Puzzle. When Loyd's smaller circle is turned, the body parts of the Warriors, like the lines in our rectangle, are realigned, making it seem that one of the Warriors, like our tenth line, has disappeared. Like optical illusions, this vanishing trick warns us of the fallibility and unreliability of the eyes in seeing what is there.

The expression "to spot a difference" comes from a related children's puzzle whereby one must spot a given number of minute differences between two otherwise similar pictures. This is known more generally as a *picture puzzle.* Solvers are given two pictures that seem identical, but which, in fact, contain tiny differences in detail. The one in Figure 4.16 is typical of this genre

Figure 4.16 Spot the differences puzzle (from Wikimedia Commons).

As trivial as this type of puzzle appears to be, it nevertheless holds an important principle related to perception. Like the vanishing tricks discussed above, it warns us, in effect, to be wary of what we see initially, for we tend not to spot tiny differences between apparently similar things. Once they are spotted, however, we tend to feel a momentary sense of satisfaction. The American lithographic company Currier & Ives featured hidden people, animals, and other objects. These held a broad appeal because they enticed viewers to detect the concealed images (Costello 1988: 51). The use of incongruous figures also has a long tradition in painting. William Hogarth and Norman Rockwell, for example, were well known for inserting incongruous figures into their paintings, thus creating a sense of ambiguity and discordance that adds considerably to the satirical effect of their art.

Rebuses

A rebus is a type of cryptographic puzzle whereby a message is enciphered by replacing words, or parts of words, with pictures, signs, other letters, layout structure, and so on. Figure 4.17 shows a rebus of the expression: "Betrayal is a stab in the back."

Figure 4.17 A rebus puzzle.

In the case of "betrayal," the image of the "bee" alludes to the pronunciation of the first two letters in the word "betrayal" (adjusted to "be"), while the image of the "tray" coincides with the "tray" in "betrayal." The suffix "-al" completes the word. For "stab," the image is a straightforward pictograph—standing directly for the concept. So, too, is the image of a "back." The idea in this type of puzzle is figuring out (literally) how the bits and pieces fit together. It thus constitutes a kind of thought model revealing how the brain may use imagery and semantics to produce whole concepts.

Rebuses based on layout play on a different aspect of cognition, namely how we read written messages, which in the case of alphabetic writing is linearly:

W1111HILE IT IS GOOD TO FEEL

ON
WORLD

The first item is read as "once in a while." This is so because there are several "ones" inserted in the word WHILE—that is, "ones in a WHILE." With an adjustment to the spelling and pronunciation we get "once in a while." Now, the word ON is on top of the word WORLD, visually representing the expression "on top of the world." So, the message, properly written, is: "Once in a while it is good to feel on top of the world."

Rebuses have been a source of fascination throughout the world and across history. It is not known where, when, or why they originated. Coins with rebuses inscribed in them, representing famous people or cities, were common in ancient Greece and Rome (Céard and Margolin 1986). During the Middle Ages, rebuses were frequently used to encode heraldic mottoes. In Renaissance Italy, Pope Paul III (1468–1549) employed them to teach writing. In the early part of the seventeenth century, the priests of the Picardy region of France put them on the pamphlets they printed for Easter celebrations, so that alphabetically illiterate people could understand parts of the message and thus reconstruct the complete message. So popular had rebuses become throughout Europe that Ben Jonson, the English playwright and poet, trenchantly ridiculed them in his play *The Alchemist*. Rebus cards appeared for the first time in 1789 (Costello 1988: 8).

Like riddles, rebuses have always been especially appealing to children. The popular nineteenth-century children's magazine titled *Our Young Folks* provided rebuses in virtually every issue (Hovanec 1978: 68). Combining pictographic with alphabetic symbolism, rebuses put on display the power of human signs to encode meaning in ingenious and imaginative ways. A pictographic representation may stand for its referent or the thing it depicts. An alphabetic representation, on the other hand, stands for the sounds in a word. But alphabetic characters still have an embedded pictographic modality in them—an element that rebuses bring out enigmatologically. Rebuses thus remind us that we encode a large part of our knowledge about the world through visual symbols, not just written words, actually using both to make sense of things.

Geometric puzzles

Geometry starts out as both measurement and as a sacred craft believed to reveal hidden secrets about reality. An example of how these two perceptions converge psychologically is the number π. Knowledge of π is ancient. It is mentioned, for instance, in the Old Testament (II Chron. 4:2), where we read: "Also he made a molten sea of ten cubits from brim to brim, round in compass and five cubits the height thereof, and a line of thirty cubits did compass it about." This tells us that the Hebrews took the ratio to be three. The Babylonians also thought it was three, and the Egyptians estimated it to be 3.1604.

Archimedes showed that it was between 3.14 and 3.142. He did this with a simple, but ingenious, method, inscribing an infinite number of polygons in a circle. The perimeter of a polygon can easily be measured (with basic geometry). Archimedes determined the perimeters of successively regular polygons of twenty-four, forty-eight, and ninety-six sides. This was based on the insight that as the number of sides in an inscribed polygon increases, the perimeter gets closer in value to the circumference of the circle, leading to π. The perimeter of a polygon with ninety-six sides falls short of the actual value of π, but for the purpose of getting an estimation it is good enough, turning out to be between 3 10/71 and 3 1/7. Six hundred years after Archimedes, around 400 CE, in a set of Indian manuscripts called *Siddhantas* ("Systems of Astronomy"), the value for π is given as 3.1416, having been estimated with a similar type of reasoning to that of Archimedes. The Chinese mathematicians also searched for a value for π. Lin Hui (*c.* 250 CE) inscribed a polygon of 3,072 sides to determine π as equal to 3.14159. All these coincidental methods for determining the value of π suggest the influence of an archetype—a "squaring the circle" archetype, which is a way to tame the structure of the circle to a measurable size. The word "squaring" is of course a misnomer in the case of π, since it is really more accurate to say "polygoning the circle."

The discovery of π has led to other fascinating serendipitous discoveries. Take, for example, a piece of cardboard and a needle. Mark parallel lines on the cardboard, spacing them the length of the needle apart. Toss the needle in the air so that it falls on the cardboard, marking its position. After a number of tosses, the scenario might look like the one in Figure 4.18.

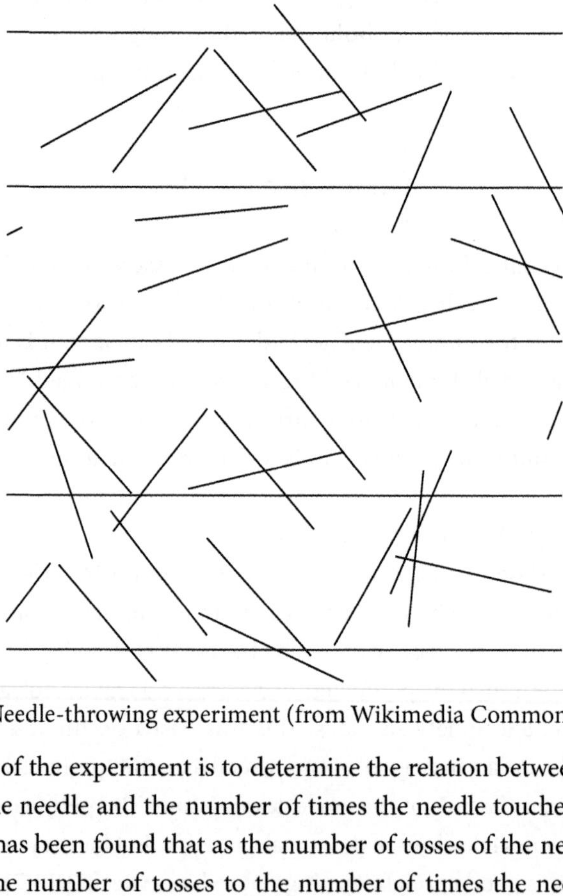

Figure 4.18 Needle-throwing experiment (from Wikimedia Commons).

The object of the experiment is to determine the relation between the number of tosses of the needle and the number of times the needle touches a line on the cardboard. It has been found that as the number of tosses of the needle increases, the ratio of the number of tosses to the number of times the needle touches a line approaches π. This is known as *Buffon's Needle Problem*, first posed in the eighteenth century by Georges-Louis Leclerc, Comte de Buffon. Margaret Willerding (1967: 120) recounts an interesting historical anecdote whereby in 1901 "a scientist made 3408 tosses of the needle and claimed that it touched the lines 1085 times," and thus the "ratio of the number of tosses to the number of times the needle touched the lines, 3408/1085, differs from π by less than 0.1 per cent."

This is nothing short of perplexing given that one can ask: Where is the circle in this game? Since there are no definitive answers to this question, all we can do is continue to play games to see what they yield.

More than any other principle in geometry, the Pythagorean theorem has been the inspiration for innumerable ingenious puzzles. The following one, for instance, was devised by the Indian astronomer and mathematician Bhaskara (*c.* 1114–85) in his famous work, the *Lilavati* (Wells 1992: 118–19). It illustrates rather nicely how the Pythagorean theorem can be utilized to produce a puzzle:

> *A snake's hole is at the foot of a pillar which is 15 cubits high and a peacock is perched on its summit. Seeing the snake, at a distance of thrice the pillar's height, gliding toward his hole, the peacock stoops obliquely upon him. Say quickly at how many cubits from the snake's hole do they meet, both proceeding an equal distance?*

What makes this puzzle seemingly intractable at first reading is the fact that neither the peacock's oblique attack on the snake nor the location of their meeting point is immediately visualizable. In other words, the answer is seemingly beyond cognitive reach, so to speak. The key to solving this is to draw a diagram that will visually represent the information presented by the puzzle, which need not concern us here. The diagram makes it obvious that the Pythagorean theorem can be applied in order to determine the required length.

Geometrical problems reach back into antiquity, as discussed several times. Most of the famous ones were solved, often leading to new branches of the field. One of the more famous geometrical mind-bogglers of all time, known as the Four-Color Problem, falls under this rubric. Mapmakers had believed from antiquity that four colors were sufficient to color any map, so that no two contiguous regions would share a color. This commonly held premise caught the attention of mathematicians in the nineteenth century, after a young mathematician at University College London, named Francis Guthrie, formally proposed in 1852 that four colors would always be enough, although there is some evidence that Möbius had proposed it in a lecture to his students as far back as 1840. Guthrie apparently wrote about the problem to his younger brother, Frederick. The story then goes that Frederick described the problem to his own professor, the prominent British mathematician and puzzlist, Augustus De Morgan, who quickly realized that the Four-Color Problem had many important ramifications for mathematical method. Word of the problem spread quickly.

The Four-Color Problem is not a puzzle in the traditional sense, since it arose initially from the observations of mapmakers. Nevertheless, it has all the structural features of a puzzle, especially since the answer seems to be elusive. It requires a large dose of spatial reasoning not only to solve, but just to envision. It is worthwhile discussing it schematically here. In its simplest form it reads as follows:

> *What is the minimum number of tints needed to color the regions of any map distinctively? (If two regions touch at a single point, the point is not considered a common border).*

Of course, a two-region diagram will require two colors and a three-region one, three. No map has been found that requires more than four colors. The challenge is to prove that this is necessarily so, no matter what diagram or map is used. In other words, the objective is to prove that four colors are *sufficient* to color any map, no matter how many regions it has. After De Morgan made the Four-Color Problem widely known, mathematicians started in earnest trying to prove it with the traditional methods of proof. But their efforts proved to be consistently fruitless. A proof finally came from two American mathematicians, Wolfgang Haken and Kenneth Appel. It was seen at first to be peculiar because it broke with the main traditions of proof: it used an algorithm, which essentially checked to see if a map could be colored by more than four tints (Haken 1977; Haken and Appel 1977, 2002). Haken and Appel (2002: 193) themselves admitted that their proof may not be the last word on the problem: "One can never rule out the chance that a short proof of the Four-Color Theorem might some day be found, perhaps by the proverbial bright high-school student."

Let's break down the Haken-Appel proof as follows (Wilson 2002):

1. They demonstrated that no single map in a set of 1,936 maps constituted a counterexample to the Four-Color Theorem, using a computer program designed to look for counterexamples.
2. Any map that could potentially constitute a counterexample must show a portion that looks like any one of the 1,936 maps. None were found.
3. Showing (2) involved hundreds of pages of hand (non-computer) analysis; it used therefore "brute force" analysis.
4. Haken and Appel concluded that no counterexample exists because it must contain, and at the same time not contain, one of the 1,936 maps—which is a contradiction.
5. The contradiction means that there are no counterexamples—hence the proof.

The program Haken and Appel wrote was thousands of lines long and took 1,200 hours to run. Since then, mathematicians have checked the program, finding only minor and fixable problems in it (Devlin 1999: 148–76). The two mathematicians essentially showed that no exception to the Four-Color Theorem can likely ever be found. However, to this day, some mathematicians remain uncomfortable with their proof. Soon after its publication, Thomas Tymoczko (1979) encapsulated the feelings of some when he observed that, if accepted, Haken and Appel's work puts mathematics in a position to radically alter its traditional view of proof. By accepting a new way of proving the theorem, mathematics did indeed undergo a paradigm shift, as Tymoczko suggests—a shift that was due essentially to a seemingly simple puzzle based on coloring maps.

The tangram, the jigsaw, and the golden ratio

As we saw in the opening chapter, Archimedes was particularly adept at experimenting with dissection and arrangement patterns. His *loculus* is essentially a dissection puzzle in which a square is cut into fourteen pieces that are to be reassembled to form silhouettes of objects or else reorganized to produce a square diagram. Archimedes' game is lost; the version that we have today comes down to us from an Arabic manuscript titled *The Book of Archimedes on the Division of the Figure Stomaschion* (Netz and Noel 2007).

The *loculus* prefigures one of the best-known manual puzzles of all time—the *tangram*. Its origin is obscure. Some believe that the word *tangram* is of American origin, derived from a Chinese word; others believe that it originated in ancient China as a kind of sacred artifact, much like the magic square (Vorderman 1996: 130). As Slocum and Botermans (1992: 8) suggest, however, the most likely source of the tangram puzzle is the late-eighteenth-century Chinese puzzle *Ch'i Ch'io*, invented during the rule of Emperor Chia Ch'ing (1796–1820). But the mathematician Takagi (1999) disagrees. He believes that the tangram is of Japanese origin because he found a remarkable little book containing seven-piece tangram puzzles, titled *The Ingenious Pieces of Sei Shonagon*, which was published in 1742 in Japan. Whatever the real origin of the tangram, the puzzle made its way to Europe and America in the early nineteenth century, where it became very popular. It is said that even Napoleon was an avid tangram player while in exile on St. Helena. Lewis Carroll, Sam Loyd, and Henry Dudeney, among other puzzlists, created many ingenious tangram puzzles. Sam Loyd devoted an entire book to this puzzle genre, *The Eighth Book of Tan*, showing his readers how truly intriguing it was (Loyd 1952).

There are seven tangram pieces—five triangles, one square, and one parallelogram. These are constructed by cutting pieces from a square template (Figure 4.19).

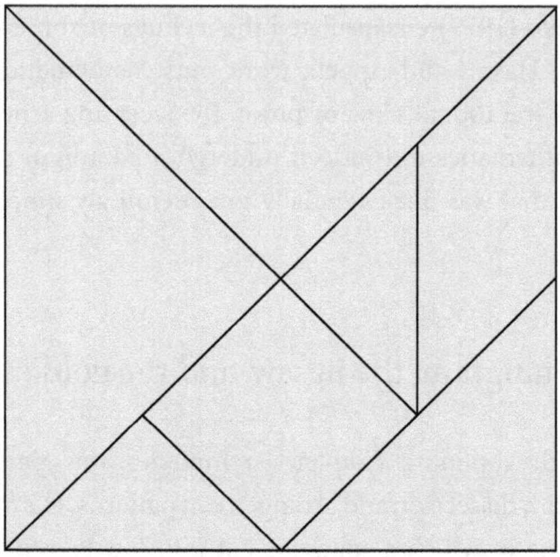

Figure 4.19 The tangram puzzle (from Wikimedia Commons).

The object is to assemble these pieces in ways to produce recognizable shapes, figures, and forms. The seven pieces may thus be cut out as many times as needed, from multiple templates, producing as many triangles, squares, and parallelograms as the puzzle calls for (Read 1965). As Loyd showed in his *Eighth Book of Tan*, the puzzle can be used to create other puzzles and paradoxes that allow us to explore hidden ideas and structures (Figure 4.20). One of these is the following:

> *The seventh and eighth figures represent the mysterious square, built with seven pieces: then with a corner clipped off, and still the same seven pieces employed.*

Figure 4.20 Loyd's tangram paradox (public domain).

Exploring this paradox is highly interesting in itself. Cognitively, it brings out the value of experimenting with a combinatory structure in order to see what it yields. It is yet another cut-and-slide illusion. Another paradox was devised by Henry Dudeney, known as the Two Monks paradox. This consists of two similar shapes, resembling "monks," which show a missing foot on one of them (Figure 4.21).

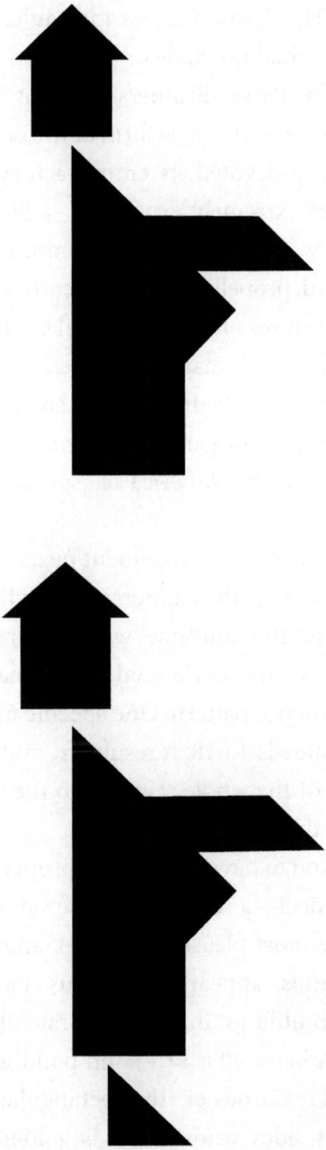

Figure 4.21 Dudeney's Two Monks paradox (from Wikimedia Commons).

This is a dissection puzzle that has a simple explanation—the area of the foot is actually compensated for by the slightly larger body in the other figure. Clearly, it is based on a similar principle to Loyd's Get Off the Earth Puzzle (Figure 4.9).

Tangrams are related to the most well-known of all dissection and arrangement puzzles—the jigsaw puzzle, which was invented by British mapmaker John Spilsbury around 1760 as a toy to educate children about geography (Hannas 1972, 1981; Williams 2004). Jigsaw puzzles for adults were put on the market around 1900. Most of these had no guide or picture on the box (unlike children's versions of the puzzles). The Parker Brothers company introduced picture guides and interlocking pieces in 1908. This new form of the puzzle became so successful that, in 1909, the company devoted its entire factory production to it. Made of wood, the puzzles were extremely expensive—a 500-piece puzzle cost $5 in 1908, when the average wage was $50 per month. By 1933, they were more cheaply made of cardboard, propelling sales to nearly 10 million per week in the United States alone. Retail stores offered free puzzles with the purchase of certain gadgets, and a twenty-five-cent magazine called *Jigsaw Puzzle of the Week* appeared on newsstands every Wednesday. Today, the jigsaw puzzle remains one of the more popular types of games, enjoyed by children and adults alike. Specialty stores throughout North America sell specialized jigsaw puzzles to suit all tastes.

Why are such dissection and rearrangement puzzles so pleasing? And why is it that some are more pleasing than others? For Pythagoras, the harmonious arrangement of things in the universe was captured by the symmetry of geometrical figures. This is why, he claimed, the human eye derives such great pleasure from seeing geometric pattern. One specific figure that he uncovered is the *golden ratio*, whose value is 1.618. It results from dividing a line segment in such a way that the ratio of the whole segment to the larger part is equal to the ratio of the larger part to the smaller part.

This ratio has been found to have astounding properties. For example, it tends to produce an aesthetic effect—a rectangle whose sides are in the ratio, tends to be perceived as being the most pleasing of all rectangles, though no one really knows why. Golden ratios appear in famous paintings, sculptures, and architectural creations. Buildings that incorporate the ratio in their design include the Parthenon (Athens, 400s BCE) and buildings designed in the 1900s by the French architect Le Corbusier. The rectangular face of the front of the Parthenon in Athens has sides whose ratio is golden. The proportion of the height of the United Nations Building in New York (built in 1952) to the length of its base is also golden (Kappraff 1991). The ratio, written as □, is $(1 + \sqrt{5}/2)$, a number mentioned at the beginning of Book VI of Euclid's *Elements*. Renaissance artists called the ratio a *divine proportion* (Huntley 1970). But the discovery of the ratio may go back in time even before Pythagoras. The ancient Egyptians, for instance, also knew of the golden ratio, which they referred to as a "sacred ratio," using it in the construction of the Great Pyramid at Giza (Pedoe 1976).

Mazes

In antiquity, labyrinths were hardly perceived to be puzzles (Chapter 1). One of the first known labyrinths was the prison built on the island of Crete, as mentioned in the opening chapter. According to legend, it was constructed by the Athenian craftsman Daedalus for King Minos to avenge the death of his son Androgeus at the hands of a group of unknown Athenians. Adding to his woe was the fact that his wife Pasiphae had fallen in love with a bull, and given birth to a half-human, half-bull beast called the Minotaur (literally, "the bull of Minos"). Embarrassed by this event, and aching to exact his revenge on Athenians generally, Minos captured and sent seven young Athenians every year into the prison. At its center he kept the voracious Minotaur, who was eager to devour anyone who ventured there. Theseus, son of King Aegeus of Athens, offered to go as one of those to be sacrificed. Ironically, Minos's clever daughter, Ariadne, had fallen in love with an Athenian—Theseus. So, she gave her beloved a sword with which to kill the Minotaur and a thread to mark his path through the labyrinth, so that he could find his way back out by retracing his steps with the path indicated by the string. Theseus slew the Minotaur and emerged to be reunited with Ariadne, finding his way back by simply following the path marked by the thread. Aegeus had instructed Theseus to raise a white sail on his ship after he had accomplished his mission. But Theseus forgot to do so and, as legend has it, when his father saw the ship returning with black sails, he threw himself into the sea, which was thereafter called the Aegean.

No one really knows what the original Cretan labyrinth actually looked like (Figure 4.22). Its most likely shape is found on ancient coins discovered at Knossos, the most probable site of the Cretan labyrinth.

Figure 4.22 Likely shape of the Cretan labyrinth (from Wikimedia Commons).

By entering at the opening, and following its single winding path, we will reach the center. This is now called a *unicursal* graph—a graph with one path through it. Mazes with alternative paths pose a much greater challenge, because there is no algorithm for solving them. However, some useful suggestions have been put forward by mathematicians over the years. The following are due to Edouard A. Lucas (1882):

1. As we go through the maze, we must constantly keep looking ahead along a path to see if it ends up being a "dead end;" if so, we must avoid it and take another one at some juncture.
2. Whenever we come to a new juncture, we look ahead to scrutinize the path as open or dead.
3. If on a path we come to an old juncture or dead end, we must turn and go back the way we came.
4. We should never enter a path marked on both sides.

The Cretan labyrinth has appealed to rulers, philosophers, mathematicians, artists, and writers alike. The later Roman emperors had copies of the labyrinth embroidered on their robes. Similar labyrinths are etched on the walls of early Christian churches. The surrealist Argentine writer Julio Cortázar was so taken by the story that he portrayed the outside world in his novels as a phantasmal Cretan maze from which every human being must escape. Cortázar's contemporary and compatriot, Jorge Luis Borges, was also spellbound by the labyrinth, writing a series of truly intriguing stories, collected under the appropriate title *Labyrinths*. The labyrinth concept constitutes an unconscious metaphor for the mystery of life, which inheres in a series of seemingly "random paths" that might conceal a "hidden pattern" leading to some "secret center" which may hold the "solution" to this mystery. In Umberto Eco's popular novel, *The Name of the Rose* (1983), the central feature is a library constructed as the Cretan labyrinth. The story tells of murders committed in a medieval monastery by a serial killer. Two clerics from outside the monastery are called in to solve the murders. When first trapped within the library of the monastery, the two sleuths are able to escape with the assistance of a thread, in an obvious allusion to the Cretan myth. At the center, they ultimately find the culprit—a blind monk who has poisoned the pages of Aristotle's book on humor. Fearful of the power of humor, the monk had taken it upon himself to eliminate all those who dared venture into the center of the maze to discover the delights and dangers of laughter.

The oldest labyrinthine design found is carved into the stone wall of a 5.000-year-old grave in Sicily. Similar carvings have been discovered throughout the world. The greatest number, over 300, has been found in Sweden and Gotland in the Baltics (Gullberg 1997: viii). Labyrinthine patterns have also been found on stone carvings in Ireland that go back to around 2000 BCE, in the Alps, at Pompeii, in Scandinavia, Wales, England, Africa and in Hopi rock carvings in Arizona (Fisher and Gerster 1990: 32–44). Many ancient buildings and cities were designed as labyrinthine structures. The Egyptian pyramids and the Christian catacombs—the networks of subterranean chambers and galleries used for burial by peoples of the ancient Mediterranean world—were designed as labyrinths, presumably to test the ability of the deceased to figure out the right path to the afterworld (Hooke 1935; Lockridge 1941).

The labyrinth concept reverberates with mystical connotations. In Java, Sumatra, and India, it has been used from time immemorial as a tool for achieving inner peace. The Navajo people in the United States see the labyrinth as a representation of how the world was created. The floors of some medieval churches had labyrinthine designs in them, to symbolize the tortuous journey of individuals toward salvation (Doob 1990: 18). One of the largest can be found at Chartres, where the faithful would make pilgrimages on their knees through it. In Renaissance Europe, labyrinths were added to murals and pavements, and many gardens were designed as mazes walled by clipped hedges. Two of the best known were built in the seventeenth century: the maze garden at Hampton Court in London; and the exquisite labyrinth in the Palace of Versailles, probably built by Louis XIV, which is adorned by thirty-nine fountains and various statues depicting characters from Aesop's Fables (Matthews 1970).

The labyrinth is clearly born in *mythos*; the maze puzzle is its ludic derivative. Complex maze puzzles started appearing after Lewis Carroll composed several ingenious ones himself, such as the one in Figure 4.23 (1958b).

Figure 4.23 One of Lewis Carroll's mazes (public domain).

The universality and antiquity of labyrinths, along with the mystique that surrounds them, supports the idea that they are archetypal, constituting models of the many "twists and turns" of human thought.

Visual imaging

Consider a famous puzzle called the *Josephus Problem*, after the Jewish historian Josephus of the first century CE, who supposedly saved his own life by coming up with the correct solution (Kraitchik 1942: 93–4). Here is a typical version of that puzzle:

> *There are fifteen Tyrants (T's) and fifteen helpless Citizens (C's) on a ship—way too many for the size of the ship. So, it is decided that the Tyrants must be thrown overboard to prevent the ship from sinking. A mythical beast, who cannot distinguish between Tyrants and Citizens, has been let loose on the ship to throw people overboard. The beast has been trained to throw over every ninth person seated in a circle. How can the Tyrants and Citizens on board be arranged in a circle so that the beast can eliminate only the Tyrants?*

The beast starts at the first "C". The ninth person from the start is a "T". So he is thrown overboard. The ninth person after that is also a "T." He, too, is thrown overboard. This guarantees that every Tyrant is thrown overboard while all the Citizens are saved. If we start going through this sequence with different numbers of people in the starting circle, we will see a few hidden patterns emerging. First, the final survivor is never someone in an even-numbered position because all of the people standing in even-numbered positions are killed first (one kills two, three kills four, and so on). Now, any time that the starting number of people is a power of two, the final person standing is the same as the person who started the sequence (position number one). This is the key to figuring out where a survivor should stand. When the number of people left standing is equal to a power of two, then it is the survivor's turn to kill a neighbor.

The original Josephus Problem consisted of a circle of forty-one men (n = 41), with every third man killed. Versions of the puzzle are found in different cultures throughout the world, implying that it is likely based on an archetype. It has been studied by famous mathematicians, including Leonhard Euler, because it constitutes, in puzzle form, a miniature model for investigating more complex problems in systematic arrangement—an area of study that now goes under the rubric of systems analysis. As Petkovic (2009: 2) states:

> This problem, appearing for the first time in Ambrose of Milan's book ca. 370, is known as the Josephus problem, and it found its way not just into later European manuscripts, but also into Arabian and Japanese books. Depending on the time and location where the particular version of the Josephus problem was raised, the survivors and victims were sailors and smugglers, Christians and Turks, sluggards and scholars, good guys and bad guys, and so on. This puzzle attracted attention of many outstanding scientists, including Euler, Tait, Vilf, Graham, and Knuth.

The puzzle is an experiment in both imagination and reasoning—a blend that can be called, simply, "visual imaging." Individual differences in the ability to experience imagery were recorded in the nineteenth century. The research has since shown that people can imagine faces and voices accurately and quickly, rotate objects in their heads, locate imaginary places, scan game boards (like a checker board) in their minds, and so on with no difficulty whatsoever. While psychologists might disagree on exactly what it is that their subjects conjure up in their minds, there is general agreement that something is "going on." Stephen M. Kosslyn (1983, 1994), who is well known for having investigated empirically how the brain's spatial reasoning system might work, has conducted a series of ingenious experiments that show how subjects can easily form images in their mind to help them carry out tasks, such as arranging furniture in a room, designing a blueprint, and so on.

The Pythagoreans were petrified by anomalies in spatial configurations. They considered their "theory of order" seriously undermined when, ironically, Pythagoras's own theorem revealed the existence of irrational numbers such as $\sqrt{2}$. This "chaotic" number stared them straight in the face each time they drew an isosceles right-angled triangle with equal sides of unit length. The length of its hypotenuse was the square root of the sum of $1^2 + 1^2$, or $\sqrt{2}$, a number that cannot be represented as the ratio of two integers, or as a finite or repeating decimal (it begins with 1.4142135 . . .). For the Pythagoreans, rational numbers had a "rightness" about them; irrational ones such as $\sqrt{2}$ did not. And yet, there they were, defying logic and sense, and challenging the system of order that the Pythagoreans so strongly desired to establish. So disturbed were they that, as some stories have it, they "suppressed their knowledge of the irrationality of $\sqrt{2}$, and went to the length of killing one of their own colleagues for having committed the sin of letting the nasty information reach an outsider" (Ogilvie 1956: 15). The colleague is suspected to have been Hipassus of Metapontum (Bunt, Jones, and Bedient 1976: 86; Aczel 2000: 19).

Incredibly, even in the domain of chaos, the human mind finds ways of making sense of things. *Chaos theory*, founded by Henri Poincaré, has shown, in fact, that there are patterns even in random events (Gleick 1987). Already in the 1960s, simplified computer models demonstrated that there was a hidden structure in the seemingly chaotic patterns of weather. When these were plotted in three dimensions, they revealed a butterfly-shaped fractal set of points. Similarly, leaves, coastlines, mounds, and other seemingly random forms produced by Nature reveal hidden fractal patterns when examined closely.

All this constitutes a truly profound existential paradox. Why is there order in chaos? Perhaps the ancient myths provide, after all, the only plausible response to this question. According to the Theogony of the Greek poet Hesiod (eighth century BCE), Chaos was produced by Earth, from which arose the starry, cloud-filled Heaven. In a later myth, Chaos was portrayed as the formless matter from which the Cosmos, or harmonious Order, was created. In both versions, the ancients felt deeply that order arose out of chaos. This feeling continues to reverberate today, and is perhaps behind the feeling of satisfaction we get when solving a geometrical puzzle or finding our way through a maze. For some truly mysterious reason, our mind requires that there be order within apparent disorder. The search for this order has been a motivating force in directing or shaping human evolution.

5

Puzzles in Mathematics

As far as the laws of mathematics refer to reality, they are not certain, and as far as they are certain, they do not refer to reality.

Albert Einstein (1879–1955)

Perhaps in no other area of human intelligence have puzzles played as large a significant role than in mathematics. One of the earliest surviving manuscripts of mathematics is, essentially, a collection of puzzles. As mentioned briefly in Chapter 1, it is referred to either as the *Ahmes Papyrus*, after the Egyptian scribe Ahmes, who copied it, or the *Rhind Papyrus*, after collector Henry Rhind, who purchased it in 1858. In addition to more than eighty-four challenging mathematical puzzles, the *Papyrus* contains tables for the calculation of areas and the conversion of fractions, elementary sequences, linear equations, and extensive information about measurement (Peet 1923; Gillings 1972; Chase 1979; Olivastro 1993; Blatner 1997).

Today, the *Papyrus* is seen as a source of ancient mathematics. It shows, significantly, that puzzles and mathematics have a common origin. So important is puzzle-making to mathematics that in 1612 "recreational mathematics" emerged as a kind of semi-autonomous branch of mathematics. In that year, Claude-Gaspar Bachet de Mézirac published one of the first comprehensive collection of mathematical puzzles, a book titled *Problèmes plaisans et délectables qui se font par les nombres*. One of the reasons why puzzles harbor so many embryonic ideas that can be shaped into theories is that they are born of the imagination, as emphasized throughout this book, and thus of archetypal thinking. Indeed, long before Archimedes came up with the proof for the value of π (Chapter 4), the same kind of proof is found in the papyrus, suggesting, as argued, that it is an archetype. The relevant puzzle in the *Ahmes Papyrus* is number 48:

What is the area of a circle inscribed in a square that is 9 units on its side?

The clever Ahmes solved it with the same Archimedean Aha insight: *What if the circle is transformed into a polygon?* He proceeded to do exactly that by trisecting each side of the square, thus producing nine smaller squares within it (each three by three). He then drew the diagonals in the corner squares. Such modifications to the diagram produced an octagon, which Ahmes assumed to be close enough in area to the circle for the practical purposes of his puzzle. The area of the octagon is equal to the areas of the five inner squares (which form the outline of a cross) plus half the areas of the four corner squares (= the area of two squares). Its area is thus equal to the sum of the areas of seven small squares. The area of one small square is three times three, or nine square units. The total area of seven such squares is, therefore, nine times seven, or sixty-three square units. With a bit of convenient cheating, the resourceful Ahmes assumed that the circle's area, which is nearly sixty-four, is effectively equal to 8^2. He then estimated the value of π as 3.16049.

The calculation of π has not been a trifling matter in the history of human civilization, as already discussed. A world in which π is not known is, of course, conceivable. But what we now know about objects in the world, like the sun and the tides, would be much more rudimentary. As Kasner and Newman (1940: 89) aptly put it, without π "our ability to describe all natural phenomena, physical, biological, chemical or statistical, would be reduced to primitive dimensions."

As Ahmes' puzzle example shows, mathematical puzzles are hardly just trivial intellectual recreations. They come from what Dehaene (1997: 151) calls the "illuminations" that mathematicians claim to see within their minds:

> They say that in their most creative moments, which some describe as "illuminations," they do not reason voluntarily, nor think in words, nor perform long formal calculations. Mathematical truth descends upon them, sometimes even during sleep.

The term *meme* was introduced by Richard Dawkins (1976) to refer to any element of a culture or system of behavior that is passed from one individual to another by nongenetic means. Dawkins' controversial notion will be used in a figurative sense in this chapter in a particular way, given its usefulness in describing the spread of ideas initially encoded in puzzles.

The classic mathematical puzzles of Alcuin, Fibonacci, and many more are examined in this chapter in the light of what they reveal about mathematical cognition and its origins in the same kind of blended imaginative-reflective thinking that undergirds all puzzles. Once a new mathematical idea is extracted from a puzzle, it becomes a kind of "intellectual meme" that makes its way into the mathematical mind to suggest new ways of doing mathematics. Alcuin's River-Crossing Puzzles led, eventually, to the concept of critical paths, and Fibonacci's Rabbit Puzzle led to the concept of recursive structure.

The Ahmes papyrus

The Ahmes Papyrus was, in all likelihood, a mathematics textbook for Egyptian students. It is divided into three main parts—problems in arithmetic and algebra, problems in geometry, and miscellaneous problems. This division itself suggests an educational design, although there is some debate as to its pedagogical uses (Danesi 2002). As mentioned previously (Chapter 1), there is no indication that the papyrus made a distinction between problems (such as measurement exercises) and puzzles, as they would be conceived today. Many of the "problems" that it contains have all the structural and ontological features of puzzles; they were likely seen as ideal for introducing Egyptian students to the type of imaginative-reflective thinking required in mathematics. As Petkovic (2009: 2) states, they suggest "that the early Egyptians based their mathematics problems in puzzle form." It was translated into German in 1877, and then into English in 1923. The first extensive edition of the work was carried out in 1929 by Arnold Buffum Chase, making the papyrus accessible for the first time to a large public (Chase 1979). It is now preserved in the permanent collection of the British Museum.

It starts off with the following enigmatic poem:

Accurate reckoning,
the entrance into the knowledge of all existing things and all obscure secrets.

Decoding the poem metaphorically, suggests that the author saw puzzles as tools for discovering hidden patterns ("obscure secrets"), which is what puzzles are all about. Even the papyrus's title—*Directions for Attaining Knowledge of All Dark Things*—speaks of something shrouded in mystery, suggesting that puzzles and secret knowledge were intrinsically intertwined in the minds of ancient peoples. Consider Problem 79, which is presented in the form of an inventory without an appurtenant question:

houses	7
cats	49
mice	343
sheaves of wheat	2,401

hekats of grain	16,807
estate	19,607

As the initial poem suggests, we are confronted with an "obscure secret;" that is, a nonobvious solution to the puzzle. The Aha insight comes when we examine the first five numbers more closely and discover that they are successive powers of 7: $7 = 7^1$, $49 = 7^2$, $343 = 7^3$, $2{,}401 = 7^4$, and $16{,}807 = 7^5$. But the last figure does not fit in with this insight. However, since it is placed at the end of the inventory, it suggests a summation. Indeed, 19,607 is the sum of these numbers: $7 + 49 + 343 + 2{,}401 + 16{,}807 = 19{,}607$.

Arguably, this puzzle conceals an archetype, based on the mystical meanings that the number seven has typically enfolded. This may be the reason why the same puzzle archetype surfaces in Fibonacci's *Liber Abaci* of 1202, who added, simply, another power of seven, namely 7^6, to his version:

Seven old women are on the road to Rome. Each woman has seven mules, each mule carries seven sacks, each sack contains seven loaves, to slice each loaf there are seven knives, and for each knife there are seven sheaths to hold it. How many are there altogether, women, mules, sacks, loaves, knives, sheaths?

Fibonacci could not possibly have known about Ahmes' puzzle, because the existence of the *Ahmes Papyrus* was not known at the time, nor had hieroglyphic writing been decoded. But the similarity between the two puzzles is unmistakable. In eighteenth-century England the same archetype appeared clothed as a popular nursery rhyme:

As I was going to St. Ives
I met a man with seven wives.
Each wife had seven sacks,
Each sack had seven cats,
Each cat had seven kits.
Kits, cats, sacks, wives,
How many were going to St. Ives?

That version, however, contained a clever trap. The sly anonymous puzzlist asked how many kits, cats, sacks, and wives were *going to* St. Ives, not *coming from* it. Only one person was *going to* St. Ives—the narrator of the rhyme. All the others were, of course, making their way *out of* the city (by implication). As these serendipitous coincidences with Ahmes' puzzle show, archetypes are likely the real "obscure secrets" that we seek to illuminate with our "ludic brain." Of course, the possibility of imitation exists. But the archeological evidence argues against this, as discussed, especially the fact that Egyptian hieroglyphs were not deciphered in Fibonacci's era.

As texts such as the Ahmes Papyrus suggest, mystery, intelligence, and puzzle-solving were intrinsically intertwined in the ancient world. Many of the first books of wisdom turn out, upon closer scrutiny, to be elaborate puzzle creations. The *I Ching*, for instance, which was written during the Shang dynasty of ancient China (*c.* 1766–*c.* 1027 BCE) and which has traditionally been used for divination, is organized around sixty-four symbolic hexagrams that are miniature puzzles in combinatorics, characterized as "cryptic poems." The *I Ching* uses two basic lines—a *yin* (broken line) and a *yang* (unbroken line)—to draw up the symbols. The lines are converted into numbers and then into symbolic answers to spiritual questions.

The number seven has a strange mystic appeal to people across the world. There are seven gods of good fortune in Japanese lore; seven chieftains in Greek mythology who undertook an ill-fated expedition against the city of Thebes; seven deadly sins, according to medieval theologians; seven cosmic truths according to the Plains peoples of North America; and the list could go on and on (Chevalier and Gheerbrant 1994: 859–68). The reappearance of Ahmes' puzzle can thus be best explained as resulting from the "unconscious symbolic appeal" that the number seven seems archetypally to possess. On the other hand, Gillings (1972: 168) offers a more mundane reason for Ahmes' use of the number seven: "The number 7 often presents itself in Egyptian multiplication because, by regular doubling, the first three multipliers are *always* 1, 2, 4, which add to 7." But, as Maor (1998: 13) remarks, this explanation is unconvincing because "it would equally apply to 3 (= 1 + 2), to 15 (= 1 + 2 + 4 + 8), and, in fact, to all integers of the form 2^n–1." The creation of puzzles with such "numerological symbolism" was not unusual in the ancient world. It characterized the emerging discipline of mathematics as emanating from both *mythos* and *lógos* entwined in a creative cognitive blend.

The magic square

The lack of distinction between numerology, or the mythical connotations of numbers, and numeration, or the use of numbers to describe the world quantitatively is noticeable in a truly remarkable ancient artifact that may constitute the world's first true mathematical puzzle—the *magic square*, called originally *Lo-Shu* in Chinese. One version of the story of Lo-Shu tells that in ancient China there was a huge flood. To calm his anger, the people offered sacrifices to the god of the Lo River. However, the only thing that happened after each sacrifice was the appearance of a turtle from the river, which walked around nonchalantly. The people saw the turtle as a sign from the god who, they thought, kept on rejecting their sacrifices, until one time a child noticed a square on the shell of the turtle. In it were the first nine digits arranged in rows and columns. The child also realized that the numbers along the rows, columns and two diagonals added up consistently to fifteen. From this, the people understood the number of sacrifices required of them before their god would be appeased. Another version of the Lo Shu legend has Emperor Yu the Great walking along the banks of the Lo River, when he saw a mysterious turtle crawl from the river. On its shell was a square arrangement of the first nine integers. Like the child, Yu noticed that the numbers in the square formed a pattern, interpreting it as a coded message from the river god.

Whichever legend is the correct one (if either), Lo-Shu was the first magic square made up of the first nine whole numbers, $\{1, 2, 3, 4, 5, 6, 7, 8, 9\}$, distributed in such a way that the three rows, three columns, and two diagonals added up to 15, known as the *magic constant* (Figure 5.1).

Lo-Shu

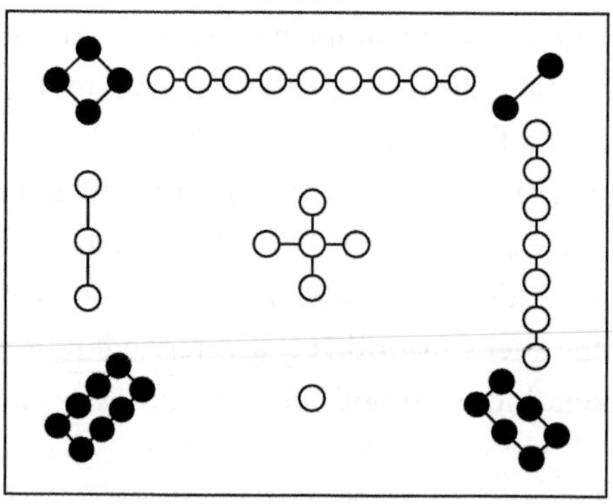

In Decimal Numbers

4	9	2
3	5	7
8	1	6

Figure 5.1 Lo-Shu (from Wikimedia Commons).

The symbolism does not stop at this arrangement. Inside the square, if we trace an S-shaped path through the even numbers, considered to be feminine, and a different S-shaped path through the odd numbers, considered to be masculine, when united, the two S-figures form the symbol for the yin-yang opposition mentioned above (Figure 5.2).

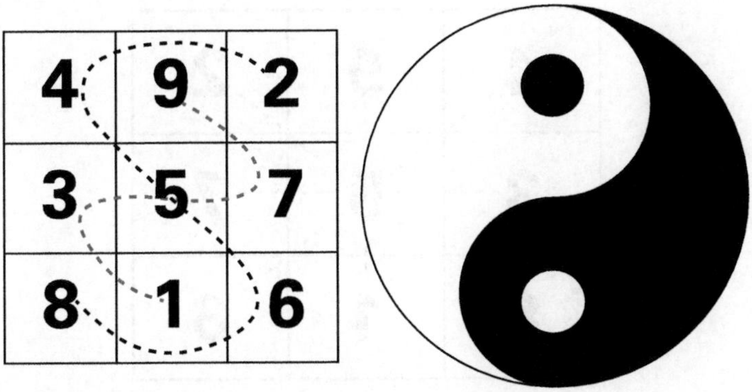

Figure 5.2 The Yin-Yang opposition in Lo-Shu (public domain).

Lo-Shu spread from China to other parts of the world in the second century CE. Devising different kinds of magic squares quickly became part of occultist traditions. Like the Chinese, medieval astrologers perceived arcane divinatory properties in them, using them to cast horoscopes. The eminent astrologer Cornelius Agrippa, for example, believed that a magic square of one cell (a square containing the single digit one) represented the eternal perfection of God. Agrippa also took the fact that a two by two magic square cannot be constructed to be proof of the imperfection of the four elements: air, earth, fire, and water.

There are many mathematical features that can be gleaned from magic squares, but a very important one concerns the notion of algorithm. Is there a method to the construction of magic squares? Or is it just a matter of trial and error? Lo-Shu is made from the first nine consecutive integers arranged into a square pattern. The last integer in the series, 9, is thus "n^2". Similarly, in an order "4" magic square, the last number is "4^2" (= 16); in an order "5" magic square, it is "5^2" (= 25); and so on. Using the summation technique for sequences, we can now set up an appropriate formula for the sum of the numbers in a magic square:

Sum of "n" numbers: $$\frac{n(n+1)}{2}$$

Note that Lo-Shu is an odd number square, that is, it is constructed with an odd number of integers. All such squares have a middle cell. And the number that fills that cell can be determined by figuring out how many rows, columns, and diagonals it occurs within the square. In the case of Lo-Shu, it occurs in one row, one column, and the two diagonals (four in total) (Figure 5.3).

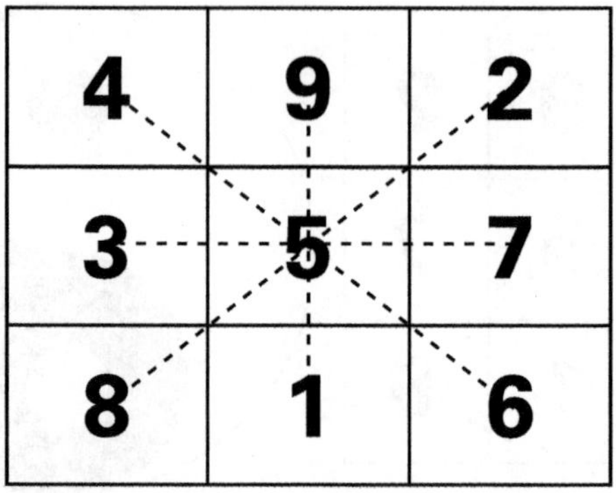

Figure 5.3 Middle cell in Lo-Shu.

It can also be identified in the following manner, since there are eight possible number triplets (made up with the first nine integers) that add up to fifteen. These will therefore make up the rows, columns, and diagonals of the square:

$9 + 5 + 1 = 15$

$9 + 4 + 2 = 15$

$8 + 6 + 1 = 15$

$8 + 5 + 2 = 15$

$8 + 4 + 3 = 15$

$7 + 6 + 2 = 15$

$7 + 5 + 3 = 15$

$6 + 5 + 4 = 15$

We established above that the middle number appears in four such triplets. Eliminating the others, we are left with:

$9 + \underline{5} + 1 = 15$

$8 + \underline{5} + 2 = 15$

$7 + \underline{5} + 3 = 15$

$6 + \underline{5} + 4 = 15$

In this way, we have identified the middle number. A similar line of reasoning can be applied to magic squares of increasing odd order. The puzzle becomes, in this way, a source for studying arithmetical structure and how it can be fleshed out of arrangements. Because of hidden structure, the study of magic squares has had an impact on the development of the concept of algorithm. Figure 5.4 is a well-known algorithm for a fourth-order magic square. First, we draw intersecting lines through the diagonals.

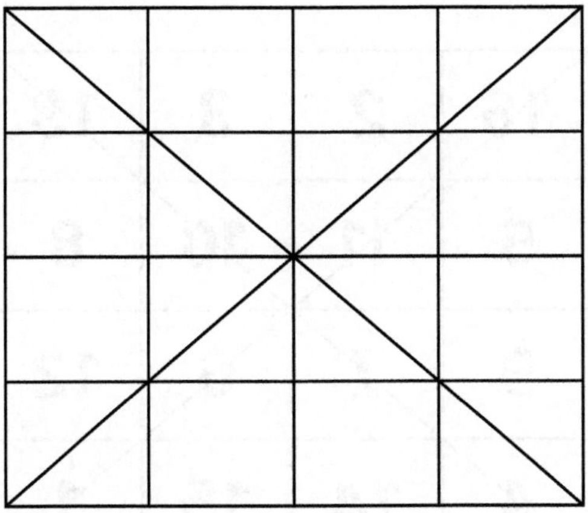

Figure 5.4 Magic square algorithm—part 1.

Next, we put in the numbers into the cells, as if they were consecutive (1, 2, 3, 4, 5, . . ., 16), leaving blank those that are crossed out by the intersecting lines. We start with 1 in the upper left corner cell. Since it is crossed, we leave it blank. We pass on to the next one to the right. Since it is empty, we put the next number in it, 2. The third cell is also empty, so we put 3 in it. The fourth cell is crossed, so we leave it empty. We proceed in this fashion until we reach the last cell in the bottom right hand corner (Figure 5.5).

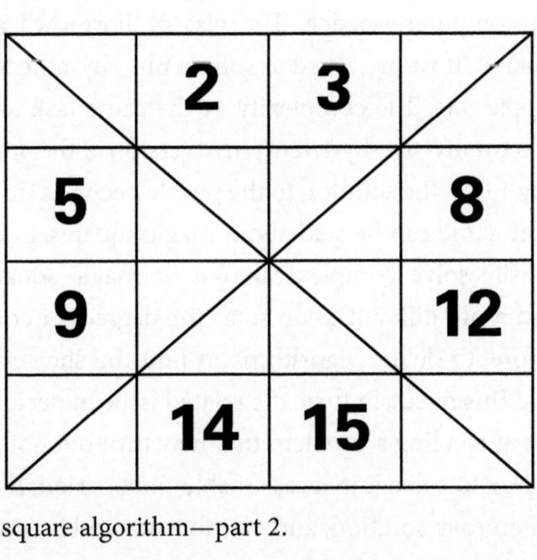

Figure 5.5 Magic square algorithm—part 2.

Now, we begin at the lower right corner, and move across the rows leftward, recording only the numbers in the cells cut by the diagonal lines. So, we start by putting 1 in the right hand corner. The next two are filled. When we reach the bottom left corner, we put in the next number, which is 4, since 2 and 3 have already been used. We continue in this way to complete the square (Figure 5.6).

16	2	3	13
5	11	10	8
9	7	6	12
4	14	15	1

Figure 5.6 Magic square algorithm—part 3.

This same kind of algorithm can be used to generate an order eight magic square. Another algorithm for constructing an odd order square is attributed to the mathematician Simon de la Loubère in 1693, although he probably learned about it during his travels to Asia. Not all magic squares yield algorithmic structure. For this reason, they have had implications for the larger question of decidability, an area that, as Fortnow (2013) has cogently argued, is at the key juncture of mathematics and computer science. The gist of Fortnow's argument can be summarized as follows. If we are asked to solve a nine-by-nine Sudoku puzzle, the task is a fairly simple one. The complexity of the same task arises when we are asked to solve, say, a twenty-five-by-twenty-five version of the puzzle. By increasing the grid to 1,000 by 1,000, the solution to the puzzle becomes immense in terms of effort and time. The same can be said about magic squares, of course. Computer algorithms can easily solve complex Sudoku or magic square puzzles, but it becomes more and more difficult to do so as the degrees of complexity increase. The idea is, therefore, to devise algorithms to find the shortest route to solving complex problems. This raises, in turn, the related issue of decidability, since there would be no point in tackling a problem that may turn out not to have a solution or else is too complex to solve it in a reasonable time period. If we let P stand for any problem with an easy solution, and NP for any problem with a difficult (or non-existent) solution, then the whole question of decidability can be represented in a simple way. If P were equal to NP, P = NP, then problems that are complex (involving large amounts of data) could be tackled easily as the algorithms become more efficient. The P = NP problem is the most important open one in computer science and formal mathematics. It seeks to determine whether every problem whose solution can be quickly checked by computer can also be quickly solved by computer. Work on this problem has made it evident that a computer would take hundreds of years to solve some NP questions and sometimes go into a loop.

Magic squares show, historically, why the early histories of mathematics and mysticism overlap. In their origins, both sought to do the same thing—unravel hidden patterns, or "obscure secrets" to cite Ahmes once again. In antiquity, no distinction was made between numeration and numerology. Numerology started with the Pythagoreans, who taught that numbers were the language of the cosmos (as discussed previously). The ancient Israelites held a similar belief, establishing the art of *gematria* on the view that the letters of any word or name could be translated into digits and rearranged to form a number that contained a secret message. The earliest recorded use of gematria, actually, was by King Sargon II of Babylon in the eighth century BCE, who built the wall of the city of Khorsbad exactly 16,283 cubits long because this was the numerological value of his name. Magic squares fall into the same category—they start off as magical artifacts, part of numerology. Over time, mathematicians transformed them into devices for studying numeration.

One of the most famous of all order-4 magic squares was constructed by the great German painter Albrecht Dürer, which he included in his famous 1514 engraving *Melancholia*. The magic square in the painting is shown in Figure 5.7.

16	3	2	13
5	10	11	8
9	6	7	12
4	15	14	1

Figure 5.7 Dürer's magic square.

Dürer's square has many hidden "magical" properties. For example, in addition to appearing in each row, column, and diagonal, the magic square constant of 34 appears as well in the following locations:

1. In the sum of the digits in the four corners (16 + 13 + 4 + 1 = 34).
2. In the sum of the four digits in the center (10 + 11 + 6 + 7 = 34).
3. In the sum of the digits 15 and 14 in the bottom row and the digits 3 and 2 facing them in the top row (15 + 14 + 3 + 2 = 34).
4. In the sum of the digits 12 and 8 in the right-hand column and 9 and 5 facing them in the left-hand column (12 + 8 + 9 + 5 = 34).
5. In the sum of the digits of each of the four squares in the corners (16 + 3 + 5 + 10 = 34; 2 + 13 + 11 + 8 = 34; 9 + 6 + 4 + 15 = 34; 7 + 12 + 14 + 1 = 34).

One of the most extraordinary of all magic squares was the order-8 magic square devised by Benjamin Franklin (Figure 5.8).

52	61	4	13	20	29	36	45
14	3	62	51	46	35	30	19
53	60	5	12	21	28	37	44
11	6	59	54	43	38	27	22
55	58	7	10	23	26	39	42
9	8	57	56	41	40	25	24
50	63	2	15	18	31	34	47
16	1	64	49	48	33	32	17

Figure 5.8 Franklin's magic square.

Franklin's square contains a host of astonishing numerical patterns, such as the following:

1. Its magic-square constant is 260; and exactly half this number, 130, is the magic square constant of each of the four 4 × 4 squares that are quadrants of the larger square.
2. The sum of any four numbers equidistant from the center is also 130.
3. The sum of the numbers in the four corners plus the sum of the four center numbers is 260.
4. The sum of the four numbers forming any little 2 × 2 square within the main square is 130. There are many more.

Magic squares harbor a philosophical impulse, which Rucker (1987: 74) describes as the "basic notion that the world is a magical pattern of small numbers arranged in simple patterns." Magic squares have developed into games today that have the same kind of recreational value of other games. They have also spawned derivative games, such as magic cubes, which are constructed with the numbers arranged in cubical form so that each row of numbers running parallel with any of the edges, and also with any of the four great diagonals, will have the same magic constant (Andrews 1960; Benson and Jacoby 1981; Pickover 2002).

Alcuin's *propositiones*

The use of puzzles in mathematics education extended to the medieval period. Early in the period, Metrodorus' *Greek Anthology* (Chapter 1) contained many of the same puzzle archetypes found in the Ahmes Papyrus. Here is one of its puzzles (Wells 1992: 23):

> *I desire my two sons to receive the thousand staters of which I am possessed, but let the fifth part of the legitimate one's share exceed by ten the fourth part of what falls to the illegitimate one.*

Using modern algebraic notation, this puzzle can be solved easily, but one cannot help but wonder how students in the sixth century CE went about solving it, when such notation was unknown. A contemporary solution can be elaborated easily as follows. We start by letting x stand for the amount of staters given to the legitimate son. A stater was an ancient coin used in various parts of Greece and elsewhere. From this, it follows that (1,000-x) represents what was given to the illegitimate son. Now, the puzzle states that the fifth part of the legitimate son's share, or $1/5x$, will exceed the fourth part of the illegitimate son's share, or 1/4(1,000-x), by ten staters. This translates into the following equation:

$$1/5x-1/4(1{,}000-x) = 10$$

The solution is $x = 577$ and 7/9 staters. This is what the legitimate son received; the other son thus got (1,000–577 7/9), or 422 and 2/9 staters. Incidentally, as is the case with many other puzzles from antiquity and the medieval period, the content of this puzzle reveals aspects of the culture of the era in which it was devised. In this case, there appears to be a religious moral subtext—legitimate children are more meritorious of inheritance than illegitimate ones, because the latter were born out of wedlock.

The awareness of the importance of puzzles to impart mathematical knowledge and to stimulate the kind of imaginative-reflective thinking that this involved in solving math problems generally was evident in Alcuin's *Propositiones ad acuendos juvenes* (Chapter 1). Below is one of its clever puzzles—Puzzle 34:

> *A certain head of a household had 100 servants. He ordered that they be given 100 modia of corn as follows. The men should receive three modia; the women, two; and the children, half a modium. Thus how many men, women, and children were there?*

This is an example of a Diophantine puzzle in which there are more variables than there are equations. The number of solutions is thus theoretically infinite, but its unique solution can be wrested out of the two equations on the basis of the given facts. Letting m, w, and c stand for the number of men, women, and children, respectively, the statement of the puzzle can be converted into the following two equations:

(1) $m + w + c = 100$
(2) $3m + 2w + 1/2c = 100$

Equation (1) states in algebraic terms that there are 100 people in all, and (2) represents the ways in which the 100 modia are distributed; that is, "m" men will receive "3m" modia in total, "w" women "2w" modia in total, and "c" children "1/2c" modia in total, for a grand total of 100 modia. Only positive integral values of m, w, and c are meaningful here, since fractional or negative integral values would have no real-life sense—people cannot be split into fractions, nor can negative numbers represent them. Thus c, which stands for the number of children, must be divisible by two, otherwise $1/2c$ in the second equation would not yield an integer. Algebraically, this can be expressed by replacing c with $2n$, the general form of an even integer (a form that reflects the fact that any number, n, when multiplied by two will always yield an even number). Substituting $c = 2n$ into the equations above produces the following:

(3) $m + w + 2n = 100$
(4) $3m + 2w + n = 100$

Now, we can multiply equation (4) by two, yielding the following equivalent equation:

(5) $6m + 4w + 2n = 200$

We can now subtract equation (3) from (5), an operation that reduces the problem to a single equation:

(6) $5m + 3w = 100$

From this, it can be seen that:

(7) $w = (100–5m)/3$

We note that m, which stands for the number of men, must be less than 99, because, if it were assigned a value of 99 or 100, the total number of people (including women and children) would be greater than 100. So, m must be a positive integral value less than 99 which, when substituted into equation (7), $w = (100–5m)/3$, will produce a positive integral value for w. If we assign the value one to m, a fractional value for w will result. If we let $m = 2$, however, then the value of w turns out to be 30. This is a definite possibility to consider further. So, we can go back to one of the two equations above, say, (3) $m + w + 2n = 100$, and substitute $m = 2$ and $w = 30$ in it. From this, the value of n turns out to be 34. Now, since $2n = c$, it is obvious that $c = 68$. We now have the solution, since 2 men, 30 women, and 68 children add up to 100 people in total. To check that our solution is correct, we give each of the 2 men three modia, each of the 30 women two modia, and each of the 68 children half a modium. This results in a total number of a hundred modia:

2 men would receive 2×3 modia	=	6 modia
30 women would receive 30×2 modia	=	60 modia
68 children would receive $68 \times 1/2$ modia	=	34 modia
		100 modia

The above solution uses contemporary algebraic notation. Without it, there is little doubt that the medieval students had to use a lot of ingenuity and inferential thinking, in addition to hypothesis testing. Like many such puzzles, it is a *Gedankenexperiment*, a thought experiment that gets solvers to consider hypotheses and follow them through to a logical conclusion. Incidentally, the same puzzle shows up in different guises in other parts of the world at other times, suggesting that it is also an archetype that is ensconced in real-world realities that seem to transcend geographical space and time. It appears much earlier in the third century CE *Bhakshali Manuscript*, discovered in northwest India in 1881. This suggests to some that Alcuin may have borrowed some of his puzzles from other cultures. But this possibility is highly unlikely, since he would have had to know many languages in an era when foreign languages were not known broadly. Solving Diophantine puzzles requires great acumen and perseverance. Alcuin certainly knew this, realizing the difference between practical problems in calculation and puzzles as specific thought experiments. Many of Alcuin's puzzles continue to find their way, in one version or other, into contemporary collections. The most famous one is the River-Crossing Puzzle (Puzzle 18), which was discussed previously. As mentioned, it is the likely meme for critical path theory and for combinatorics, since it involves pairing items in distinctive ways.

The popularity of the *Propositiones* was matched by several other medieval compilations. One of these, a collection of a hundred mechanical puzzles titled *Kitab al-hiyal* ("The Book of Ingenious Devices"), was put together by the eighth-century inventor Mohammed ibn Musá ibn Shakir of Baghdad. Another, titled *The Book of Precious Things in the Art of Reckoning*, was written by the ninth-century Egyptian mathematician Abu Kamil. By the thirteenth century, such anthologies had become commonplace. Of those that have come down to us, Fibonacci's *Liber Abaci*, published in 1202, is undoubtedly the most well known (Fibonacci 2002).

Fibonacci's *Liber Abaci*

Alcuin's *Propositiones* are fundamentally experiments in mathematical thinking. As such, his book laid the foundation for recreational mathematics to eventually develop as a field that uses puzzles to investigate abstract mathematical structures and ideas, even though it certainly was not named as such. The second foundational book in recreational mathematics is Fibonacci's *Liber Abaci*. Fibonacci designed it, actually, as a practical introduction to the Hindu-Arabic number system, which he had learned to use during his extensive travels to the Middle East. To get the system accepted in Italy and more broadly in Europe, Fibonacci had to explain the *zero* concept that mystified the philosophers of his era. The 0 symbol probably originated as far back as 600 BCE in India, although similar symbols existed in other parts of the ancient world at different times. It was the ninth-century Persian scholar, Al-Khwarizmi, who introduced the symbol 0, which he called *as-sifr* "number emptiness" (a translation of the Hindu word *sunya* meaning "void" or "empty"), to Europe before Fibonacci. If "0" stood for "nothing," people of the era argued, then it surely was "nothing," and thus had no conceivable uses. Fibonacci solved their dilemma by showing that 0 did, indeed, have a very practical function. It was a convenient arithmetical sign—a "place-holder" for separating columns of figures. So, he started off his book reassuring readers that zero allowed for all numbers to be written (cited in Posamentier and Lehmann 2007: 11):

> The nine Indian figures are: 9 8 7 6 5 4 3 2 1. With these nine figures, and with the sign 0, which the Arabs call zephyr, any number whatsoever is written.

As discussed in Chapter 1, one of the puzzles in the book has had far-reaching implications, since its solution turns up in many ways as patterns in Nature, in art, and other areas of human invention. Known as the Rabbit Puzzle, it is perhaps the best-known puzzle of math history and is worth considering here a little more. It is found in the third section of the *Liber* and its solution was discussed in the opening chapter:

> 1, 1, 2, 3, 5, 8, 13, 21, 34, 55, 89, 144, 233

This sequence has been found to conceal many unexpected mathematical patterns or "obscure secrets," to cite Ahmes again. For example, if the nth number in the sequence is x, then every nth number after x turns out to be a multiple of x. This is just the tip of the mathematical iceberg, so to say, since its secrets extend beyond mathematics itself. The series also surfaces in the morphology of daisies, which tend to have twenty-one, thirty-four, fifty-five, or eighty-nine petals (= the eighth, ninth, tenth, and eleventh numbers in the sequence); similarly, Fibonacci numbers are found concealed in the form of trilliums, wild roses, bloodroots, columbines, lilies, and irises. A major chord in Western music is made up of the octave, third, and fifth tones of the scale, that is, of tones 3, 5, and 8 (another short stretch of consecutive Fibonacci numbers). The list of such appearances of Fibonacci numbers is truly astounding (Adam 2004; Posamentier and Lehmann 2007).

There is no evidence to suggest that Fibonacci himself was aware of the implications and applications that the solution to his puzzle would turn out to have. As mentioned in Chapter 1, its recursive pattern was identified by Albert Girard in 1632. But it was the French mathematician François Edouard Anatole Lucas who noticed its far-reaching significance in the nineteenth century. Ever since, the number of mathematical properties that Fibonacci's sequence has been found to conceal and the number of its serendipitous manifestations in Nature and human life have been absolutely astonishing. The question that this episode in mathematical history raises is an obvious one: How could such a simple puzzle, designed originally to represent the efficiency of decimal numerals over Roman ones in keeping track of counting processes, contain so many implications, applications, and serendipitous appearances in the world? The plausible answer is that it taps into some archetype of the unconscious mind that is designed to interpret incoming information about the world in ways that allow us to explore that world in return. The archetype here involves a branching-tree structure, now called a Markov Chain, which is defined as a stochastic model describing a sequence of possible events in which each event depends on the state attained in the previous event. Tree structure is now a common design feature of computers and algorithmic processes.

As mathematicians started to see the Fibonacci numbers appear in the most unexpected places and in surprising ways, they became interested in finding an efficient method for calculating any Fibonacci number. In principle, this is not a difficult generalization problem. To identify the "100th" Fibonacci number, for instance, all we have to do is add the "98th" and "99th" numbers together. However, this still means we have to identify all the numbers up to the "98th", which can prove to be quite tedious. So, in the middle of the nineteenth century, the French mathematician Jacques Binet elaborated a formula, based on the calculations of Leonhard Euler and Abraham de Moivre beforehand, that allows us to find any Fibonacci number, if its position in the sequence is known. The Binet formula is given below:

$$F_n = \frac{1}{\sqrt{5}}\left(\left(\frac{1+\sqrt{5}}{2}\right)^n - \left(\frac{1-\sqrt{5}}{2}\right)^n\right)$$

It is beyond the purpose here to explain how Binet arrived at the formula. Suffice it to say that it relies on the golden ratio. The fact that Fibonacci's sequence and the golden ratio seem to form some kind of conceptual blend is itself an amazing fact. It certainly is one of Ahmes' obscure secrets that still needs to be unraveled.

Recreational mathematics

After Fibonacci, mathematicians started seeing puzzles more and more as devices for "thinking mathematically," not just recreational artifacts. However, the occasional knack to use trickery to formulate puzzles did not disappear, and much like the ancient riddlers did, mathematicians were as much involved in producing a Gotcha Effect as an Aha one. The puzzle below, from the pen of Renaissance mathematician Niccolò Tartaglia, is a case in point:

A man dies, leaving 17 camels to be divided among his heirs, in the proportions 1/2, 1/3, 1/9. How can this be done?

Dividing up the camels in the manner decreed by the father would entail having to split up a camel. This would, of course, kill it. So, Tartaglia suggested, "borrowing an extra camel," for the sake of mathematical argument, not to mention humane reasons. With eighteen camels, we arrive at a practical solution: one heir was given 1/2 (of 18), or 9, another 1/3 (of 18), or 6, and the last one 1/9 (of 18), or 2. The 9 + 6 + 2 camels given out in this way, add up to the original seventeen. The extra camel could then be returned to its owner. The clever Tartaglia devised his puzzle as a play on the often artificial connection between "real-life" conditions and mathematics. Nonetheless, as Petkovic (2009: 24) observes, the puzzle bears mathematical implications, as Tartaglia himself maintained, finding solutions to the n-camel version of the puzzle. If there are three brothers, a, b, and c, and the proportions are 1/a: 1/b: 1/c, then these are solvable by the following Diophantine equation:

$$n / (n + 1) = 1/a + 1/b + 1/c$$

This produces the following solutions:

$n = 7$ ($a = 2, b = 4, c = 8$)
$n = 11$ ($a = 2, b = 4, c = 6$)
$n = 11$ ($a = 2, b = 3, c = 12$)
$n = 17$ ($a = 2, b = 3, c = 9$)
$n = 19$ ($a = 2, b = 4, c = 5$)
$n = 23$ ($a = 2, b = 3, c = 8$)
$n = 41$ ($a = 2, b = 3, c = 7$)

As recreational mathematics started to become more and more of a mode of inquiry in Tartaglia's era, many other mathematicians became involved in puzzle creation. Consider a classic puzzle, devised by Bachet—a puzzle that he included in his 1612 collection, mentioned briefly in Chapter 1 (Bachet 1984):

What is the least number of weights that can be used on a scale to weigh any whole number of pounds of sugar from 1 to 40 inclusive, if the weights can be placed on either of the scale pans?

We might, at first, assume that all the weights need to be used. The reasoning might go somewhat as follows:

1. Weigh "1 pound" of sugar by putting the "1-pound weight" on the left pan, pouring sugar on the right pan until the pans balance.
2. Weigh "2 pounds" of sugar by putting the "2-pound" weight on the left pan, pouring sugar on the right pan until the pans balance.
3. Weigh "3 pounds" of sugar by putting the "3-pound" weight, or equivalently, the "1-pound" and the "2-pound" weights on the left pan, pouring sugar on the right pan until the pans balance.
4. In this way, we could weigh any number of integral (whole-number) pounds of sugar from "1 pound" to "40 pounds."

Clearly, the task would be cumbersome and trivial mathematically. Bachet's puzzle is hardly that, because it requires us to use "the least number of weights," not all of them. Since we can put the weights on both pans of the scale, the whole task can be done ingeniously with only four weights—the "1", "3", "9", and "27" pound weights. The reason for this is remarkably simple, but non-obvious—placing a weight on the right pan, along with the sugar, is equivalent to taking its weight away from the total weight on the left pan. For example, if "2 pounds" of sugar are to be weighed, we would put the "3-pound" weight on the left pan and the "1-pound" weight on the right pan. The result is that there are "2 pounds" less on the right pan. We will get a balance when we pour the missing "2 pounds" of sugar on the right pan.

Now, the solution is just the start. There are various mathematical properties in the solution that are interesting in themselves and that can be unpacked through reasoning. The four weights are, upon closer scrutiny, powers of "3":

$1 = 3^0$
$3 = 3^1$
$9 = 3^2$
$27 = 3^3$

These weights are sufficient because the whole numbers from "1" to "40" (= the required weights) turn out to be either a multiple or power of "3", or else, one more or less. Thus, each of the first forty integers can be expressed with the above four powers via addition and subtraction—mirroring the weighing system just described:

1	=	3^0	(= 1)
2	=	$3^1 - 3^0$	(= 3 − 1)
3	=	3^1	(= 3)
4	=	$3^1 + 3^0$	(= 3 + 1)
5	=	$3^2 - 3^1 - 3^0 = 3^2 - (3^1 + 3^0)$	(= 9 − (3 + 1))
...			
40	=	$3^3 + 3^2 + 3^1 + 3^0$	(= 27 + 9 + 3 + 1)

All we have to do is "translate" addition in the layout above as the action of putting weights on the left pan and subtraction as the action of putting weights on the right pan (along with the sugar), as shown in Table 5.1.

Table 5.1 Bachet's puzzle in chart form

Amount of Sugar to Be Weighed	Weight to be Placed on the Left Pan	Weight Added to the Right Pan along with the Sugar
1	3^0 (= 1)	none
2	3^1 (= 3)	3^0 (= 1)
3	3^1 (= 3)	none
4	$3^1 + 3^0$ (= 3 + 1)	none
5	3^2 (= 9)	$3^1 + 3^0$ (= 4)
...	...	...
40	$3^3 + 3^2 + 3^1 + 3^0$ (= 27 + 9 + 3 + 1)	none

The puzzle thus reveals something that might escape attention about integers and the operations of addition and subtraction. It is an exemplification of a principle discussed throughout this book, namely that a simple puzzle packs nonobvious information within it in an imaginative way that allows us to unpack it in an ingenious way. In effect, it casts light on an abstract structural aspect of numeration that may not have been obvious beforehand.

Another widely known recreational mathematics anthology of the era was Henry van Etten's *Mathematical Recreations, or, A Collection of Sundrie Excellent Problemes out of Ancient and Modern Phylosophers Both Usefull and Recreative*, published in French in 1624 and then in English in 1633. The title is likely the first time that the term "mathematical recreations" is used. Van Etten borrowed freely from the work of his predecessors, especially from Bachet and the *Greek Anthology*; but he also introduced many innovative puzzles of his own. Shortly thereafter, in 1647, the first collection of puzzles in America, modeled after Bachet's and Van Etten's compilations, was published in an almanac printed by Samuel Danforth, an emigré from England.

In the eighteenth century, an increasing number of mathematicians, professional and amateur, created puzzles as a means of stimulating broad interest in new ideas and as the basis for illustrating them didactically. Leonhard Euler, as we saw (Chapter 1), produced the Königsberg's Bridges Puzzle, which showed that it is impossible to trace a path over a network if it did not possess certain properties. Several decades after Euler's demonstration, mathematicians began studying paths and networks seriously. Their efforts led to the establishment of topology, which explores the properties of all kinds of networks, as well as shapes and configurations. With his Thirty-Six Officers Puzzle of 1779, Euler was also able to arouse significant interest in what was then a fledgling and uninviting area of mathematics—combinatorics. The puzzle has no known solution.

A derivative of Euler's puzzle, known as *Kirkman's School Girl Puzzle*—named after the notable amateur mathematician Thomas Penyngton Kirkman, who posed it in 1847—did have solutions and important implications for matrix theory. A matrix is a rectangular array of numerical or algebraic symbols arranged in columns and rows:

How can 15 girls walk in 5 rows of 3 each for 7 days so that no girl walks with any other girl in the same triplet more than once?

Solutions involve arranging fifteen numerals (each one representing a specific girl) in five rows of three each within seven sets (each set corresponding to a day of the week) so that no three specific numerals appear in a row more than once. It is a solvable version of Euler's officers puzzle above—suggesting that we are dealing with yet another puzzle archetype here. It can also be argued that it is a prototype or meme that has made its way into Sudoku. Figure 5.9 shows one possible solution to Kirkman's puzzle, showing how the numerals, numbered from zero to fourteen, can be arranged according to the puzzle conditions (adapted from Ball 1972: 287):

Since 1922, various solutions to Kirkman's problem have been found (Gardner 1997: 125–6). Puzzles such as this interested mathematicians throughout the nineteenth century. Augustus De Morgan, for example, produced a truly ingenious work in recreational mathematics, titled *A Budget of Paradoxes*, in which he explored a host of mathematical theories, ideas, and suppositions that were being bandied about at the time. Lewis Carroll also produced two major works in recreational mathematics—*Pillow Problems* in 1880 (seventy-two puzzles in arithmetic, algebra, geometry, trigonometry, calculus, and probability) and *A Tangled Tale* in 1886 (puzzles originally published in monthly magazine articles). Carroll was an eminent mathematical theorist, writing two famous

Monday		
0	5	10
1	6	11
2	7	12
3	8	13
4	9	14

Tuesday		
0	1	4
2	3	6
7	8	11
9	10	13
12	14	5

Wednesday		
1	2	5
3	4	7
8	9	12
10	11	14
13	0	6

Thursday		
4	5	8
6	7	10
11	12	0
13	14	2
1	3	9

Friday		
4	6	12
5	7	13
8	10	1
9	11	2
14	0	3

Saturday		
10	12	3
11	13	4
14	1	7
0	2	8
5	6	9

Sunday		
2	4	10
3	5	11
6	8	14
7	9	0
12	13	1

Figure 5.9 A solution to Kirkman's puzzle.

treatises, *A Syllabus of Plane Algebraical Geometry* (1860) and *Euclid and His Modern Rivals* (1879), both of which were widely read by the mathematicians of the era.

One of the most famous puzzles of Carroll's era, called the Towers of Hanoi, was invented by Edouard Lucas, which he published in 1883 under the pseudonym M. Claus de Siam, an anagram of his name. It is paraphrased below:

> *A monastery in Hanoi has three pegs. One holds 64 gold discs in descending order of size—the largest at the bottom, the smallest at the top. The monks have orders from God to move all the discs to the third peg while keeping them in descending order. A larger disc must never sit on a smaller one. All three pegs can be used. When the monks move the last disk, the world will end. Why?*

The puzzle starts with the discs neatly stacked in order of size on the left-most peg, smallest at the top, thus making a conical shape. The object of the game is to move the entire stack to the last peg to the right, according to the following two simple rules: (1) only one disc may be moved at a time; (2) no disc may be placed on top of a smaller disc during any move and thus on any peg. The solution reveals a recursive structure. The original version had sixty-four discs and three pegs. It would take $2^{64}-1$ moves to accomplish the task of moving the discs as stipulated, from the left-most peg to the right-most one. Even at one move per second (with no room for mistakes), the task would require 5.82×10^{11}, or 582,000,000,000 years to accomplish.

Figure 5.10 shows a three-disc version of the puzzle.

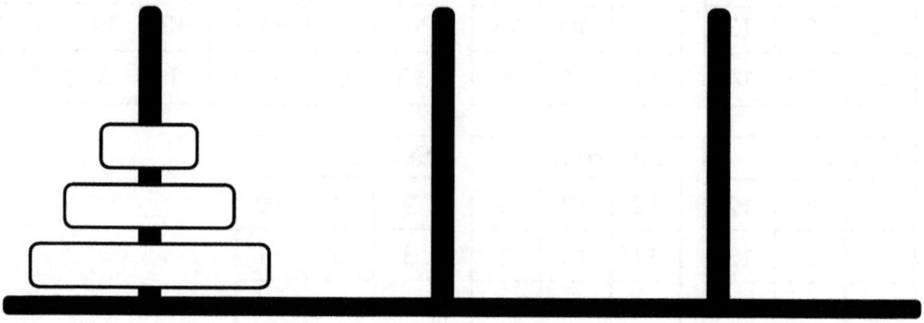

Figure 5.10 Simpler version of the Towers of Hanoi puzzle (from Wikimedia Commons).

To keep track of the moves, it is useful to number the discs and name the pegs A, B, and C from left to right (Figure 5.11). The moves are as follows:

1. Move disc 1 from A to C
2. Move disc 2 from A to B
3. Move disc 1 from C to B on top of 2
4. Move disc 3 from A to C
5. Move disc 1 from B to A
6. Move disc 2 from B to C on top of 3
7. Move disc 1 from A to C on top of 2 which is itself on top of 3

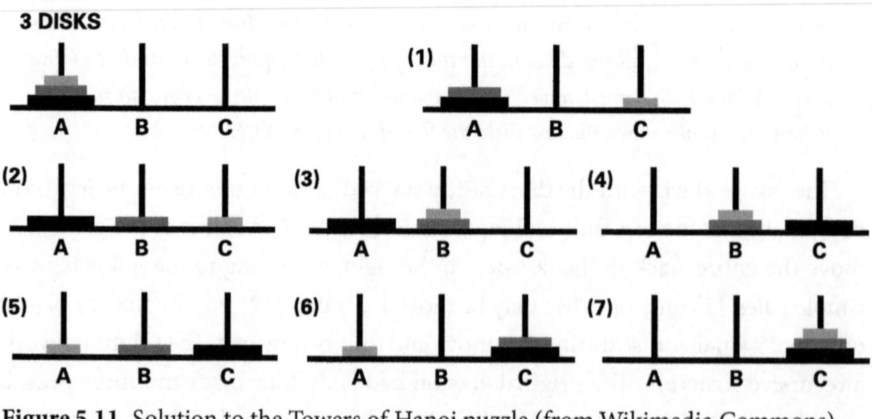

Figure 5.11 Solution to the Towers of Hanoi puzzle (from Wikimedia Commons).

It took seven moves to accomplish the task. The result can be represented as 2^3-1. We note that the exponent stands for the number of discs in the game. If we try the same game out with four discs, we will find that the number of moves is 2^4-1. It is already obvious that a general pattern is involved. If we were to play the Towers of Hanoi game with four, five, and higher numbers of discs, we would, in fact, find that the number of moves increases according to the general formula 2^n-1. Table 5.2 is a summary of the first 64 versions of the game.

Table 5.2 Towers of Hanoi chart

Discs	**Number of Moves Required: 2^n-1 (n = number of discs)**						
1	2^n-1	=	2^1-1	=	(2 −1)	=	1
2	2^n-1	=	2^2-1	=	(4 −1)	=	3
3	2^n-1	=	2^3-1	=	(8−1)	=	7
4	2^n-1	=	2^4-1	=	(16−1)	=	15
5	2^n-1	=	2^5-1	=	(32−1)	=	31
6	2^n-1	=	2^6-1	=	(64−1)	=	63
7	2^n-1	=	2^7-1	=	(128−1)	=	127
...	...						
64	2^n-1	=	$2^{64}-1$	=	twenty digit number (18,446,744 ...)		

The twentieth century witnessed the establishment of recreational mathematics as an autonomous branch of the discipline. W. W. Rouse Ball's *Mathematical Recreations and Essays*, which was published originally in 1892 and reissued in at least thirteen more editions over the subsequent century, is arguably the first proclaimed textbook in the field. Martin Gardner further ensconced recreational mathematics as a serious branch of mathematics through his column for *Scientific American* that he wrote for thirty years (starting in December 1956).

Recreational mathematics was also an area exploited by both Henry Dudeney and Sam Loyd to produce truly challenging puzzles. They even invented new puzzle genres in this domain. One is called *cryptarithmetic* (Brooke (1969), generally attributed to Loyd, although this turns out to be incorrect, since the first documented appearance of a cryptarithm puzzle is in an 1864 issue of *The American Agriculturist*. So, the actual creator remains anonymous, even though Loyd became an ingenious creator of cryptarithmetical puzzles. There are actually two subgenres—*cryptarithms* and *alphametics* (Hunter 1965). A *cryptarithm* is a puzzle in which some of the digits in an arithmetical problem have been deleted. The solver is required to reconstruct the problem by deducing numerical values on the basis of the mathematical relationships indicated by the various arrangements and locations of the numbers. Cryptarithms are, in effect, the arithmetical counterparts of cryptograms. Alphametics, on the other hand, involve words and their letters, which have replaced digits.

Figure 5.12 shows the puzzle that Loyd (1959–60: 39) used to make cryptarithms staples of contemporary recreational mathematics (in the non-theoretical sense).

```
                * 5 3
          -------------
* * 9 | 6 * 8 * * *
        * * * 2
        -------
          * 9 * *
          * * 4 *
          -------
            * * 4 *
            * * * *
            -------
```

Figure 5.12 Loyd's cryptarithm puzzle.

Without discussing the details of the solution here, it is sufficient to note that the solver must reconstruct the actual problem by thinking inferentially, deconstructing the structure of division along the way. Each step in the solution requires considering reasons why certain decisions are taken and, thus, their arithmetical rationale.

Dudeney is the one who developed alphametics, in which numbers in an arithmetical layout are replaced by letters constituting actual words. Figure 5.13 shows Dudeney's original alphametic (published in the July 1924 issue of *Strand Magazine*).

```
   S E N D
 + M O R E
 ---------
 M O N E Y
```

Figure 5.13 Dudeney's alphametic.

Again, without going into a detailed discussion of its solution, it is evident that inferential reasoning and trial-and-error are involved here as well; the decoded problem is shown below:

$$\begin{array}{r} 9567 \\ +1085 \\ \hline 10652 \end{array}$$

Alphametics lay bare the structure of the mathematical operations by playing on it—trying out one thing for another and contemplating the substitutes through structural reasoning.

Mathematical method

Starting with Bachet, specific puzzles have been used by mathematicians to exemplify or experiment with mathematical ideas. Generally, puzzles make evident what the basic outline of this method is all about. Take, for example, the following puzzle:

> *I have six billiard balls, one of which weighs less than the other five. Otherwise they all look identically the same. How can I identify the one that weighs less on a balance scale with only two weighings?*

Weighing puzzles such as this one are best approached with trial runs using fewer items. This approach allows us to see if there is some general principle or procedure that can be applied to solving the original puzzle and for extending it to more complex versions. For this particular puzzle, it is best to consider first the weighing of two balls. Clearly, both balls can be put on the pans of the scale at the same time—one on the left pan and one on the right pan. The pan that goes up, of course, is the one holding the ball that weighs less. In this case, one weighing was enough to identify the culprit ball. Next, we consider the weighing of four balls. First, we divide the four balls equally in half: that is, into two sets of two balls each. We put two balls on the left pan and two on the right pan at the same time. The pan that goes up contains the ball that weighs less, but we do not yet know which one of the two. So, we take the two suspect balls from the pan that went up, discarding the ones on the other pan. We put each one of these on a separate pan—one on the left pan and one on the right pan. The pan that goes up contains the ball that weighs less.

These two trial runs have shown us how to go about methodically identifying the suspect ball in a collection of an even number of balls. We are now ready to turn our attention to the original puzzle. There are six balls in the collection, and we are told to identify the suspect ball in only two weighings. Is that possible? We start off in the same way as we did before: that is, we divide the six balls equally in half, as two sets consisting of three balls each. Then, we go ahead and perform the first weighing as before. We put three balls on each pan this time—three on the left pan and three on the right pan. The pan that goes up contains the ball that weighs less, but we do not yet know which one of the three. Now, for the second weighing, we focus our attention on the set containing the suspect ball, discarding the ones on the other pan. Is it possible to identify the suspect ball in just one more weighing (recall that the puzzle asks us to identify it in just two weighings)? We select any two of the three balls to weigh, putting the third ball aside. We put each one on a separate pan—one on the left pan and one on the right pan. What are the possible outcomes of this second weighing? If they balance, then the suspect ball is the one put aside; if one of the pans goes up, then it contains the suspect ball. These are the only two outcomes; so, one way or the other, we have identified the suspect ball.

It took just two weighings, and a little bit of clever Aha thinking, to identify the ball that weighs less. The interesting thing to note is that we penetrated the structure of the puzzle by trying out different versions. The next thing to do is to examine what it entails mathematically. In other words, can we generalize the structure of the puzzle? So, we can continue the "weighing experiment" with more and more balls in different numerical combinations in order to flesh out some general pattern and then test this against other similar patterns in other areas of mathematics. That is the essence of mathematical method. Puzzles require that we make no *a priori* assumptions and, more importantly, that we think through some procedure—such as measuring balls on a weighing scale—in a systematic way in order to extract from it some general principle. Working out a solution for simpler versions of a puzzle is thus a basic heuristic strategy. This will allow us to see if there is a general principle inherent in the simpler versions that can be utilized profitably. When a general principle is identified, we can test it out for consistency and what it tells us about its abstract structure by formulating it in some mathematical way.

Consider now the following puzzle, composed by Dudeney who first published it in his column in the *Strand Magazine* (1908) and later in his *Canterbury Puzzles* (1958):

Inside a rectangular room, measuring 30 feet in length and 12 feet in width and height, a spider is at a point, A, on the middle of one of the end walls, which is 1 foot from the ceiling; and a fly is on the opposite wall, at B, 1 foot from the floor in the centre. What is the shortest distance that the spider must crawl in order to reach the fly, which remains stationary? Of course the spider never drops or uses its web, but crawls fairly.

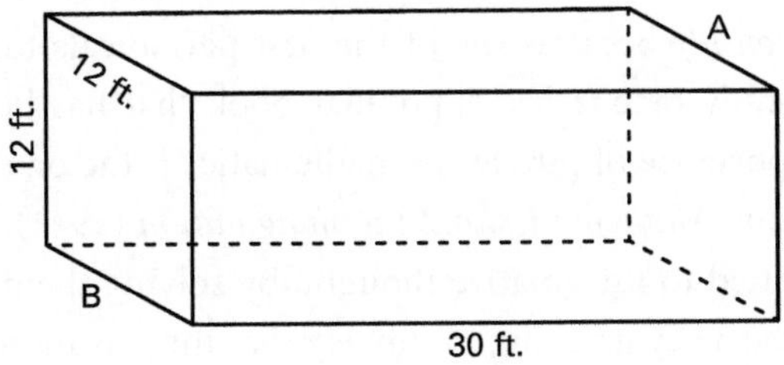

The Aha insight here is to "flatten the room" into its blueprint version (Figure 5.14).

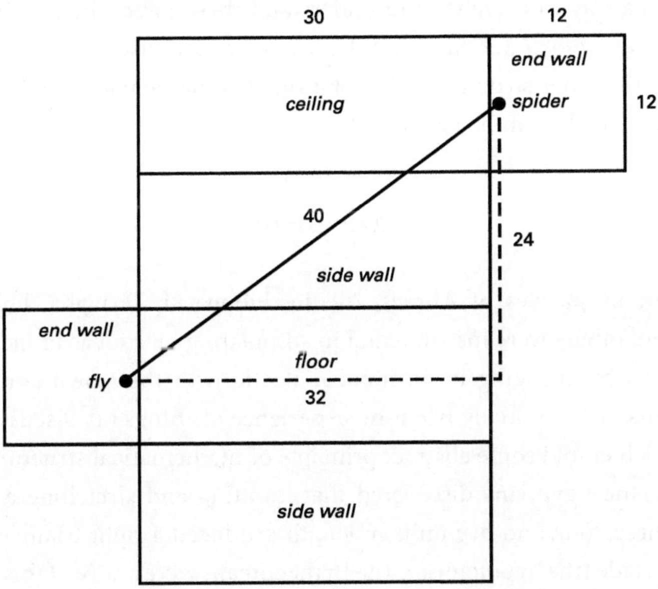

Figure 5.14 Solution to Dudeney's puzzle (from Wolfram Mathworld).

Pursuant to this insight, we can now solve this as a simple "problem" in geometry. The room has dimensions 30 × 12 × 12. The spider is in the middle of one of the 12 × 12 walls, one foot from the ceiling. The fly is in the middle of the opposite wall, one foot from the floor. The fly remains stationary, so the spider must crawl along the walls, the ceiling, and the floor in order to capture the fly on the hypotenuse of the right triangle shown above, marking the lengths as shown. The answer is 40, given that $\sqrt{(24^2 + 32^2)} = 40$.

The study of all kinds of puzzles, and their relation to mathematical method, constitutes one of the main objectives of contemporary recreational mathematics. The term *recreational mathematics* spread in the 1920s after the *American Mathematical Monthly* became one of the first periodicals to provide a venue for the formal study of puzzles. A popular book that has been influential in stressing the importance of puzzles to mathematics is the one by James Kasner and John Newman, *Mathematics and the Imagination* (1941). In it, we can see how puzzles are tied to imaginative thought. By solving them successfully and patiently, we come away grasping intuitively that there is more to mathematics than logical analysis; it is also part of the imagination and an art in itself. Another classic book is the one by Reuben Hersh, *What is Mathematics, Really?* (1998). Hersh portrays mathematics as a social phenomenon, a product of historical forces, not just an exercise in abstract cognition. Recreational mathematics constitutes a way to decipher and understand these forces. In their interesting book, *Taking Sudoku Seriously* (2011), Jason Rosenhouse and Laura Taalman show how the inner structure of Sudoku mirrors mathematical method, as will be discussed further in the next chapter.

Puzzle memes

The ingenious puzzles of Ahmes, Alcuin, Fibonacci, Tartaglia, Euler, Lucas, Carroll, and others were instrumental in stimulating new ideas in mathematics. They did so by engaging in an inner dialectic with their own constructs—a dialectic that arose initially from the experience of things (as discussed several times) which enfold some abstract principle of mathematical structure. Around 2000 BCE, the Egyptians discovered that knotting and stretching a rope into sides of three, four, and five units in length produced a right triangle, with five the longest side (the hypotenuse). The Pythagoreans were aware of this discovery. Their goal was to show that it revealed a general structural pattern. Knotting any three stretches of rope according to this pattern—for example, six, eight, and ten units—will produce a right triangle because $6^2 + 8^2 = 10^2$ $(36 + 64 = 100)$. As the historian of science Jacob Bronowski (1973: 168) has insightfully written, we hardly recognize today how important this discovery was:

> The theorem of Pythagoras remains the most important single theorem in the whole of mathematics. That seems a bold and extraordinary thing to say, yet

> it is not extravagant; because what Pythagoras established is a fundamental characterization of the space in which we move, and it is the first time that it is translated into numbers. And the exact fit of the numbers describes the exact laws that bind the universe. If space had a different symmetry the theorem would not be true.

All this suggests the presence of an "anthropic principle," as Al-Khalili (2012: 218) calls it, which implies that we are part of the world in which we live and thus privileged, in a way, to understand it best:

> The anthropic principle seems to be saying that our very existence determines certain properties of the Universe, because if they were any different we would not be here to question them.

We could conceivably live without the Pythagorean theorem. It tells us what we know practically—that a diagonal distance is shorter than taking an L-shaped path to a given point. And perhaps this is why it emerged anthropically. The River-Crossing Puzzles of Alcuin describe something that seems obvious, but whose abstract structural features escape reflection, until they are questioned. The classic math puzzles possess this feature—they unpack the hidden principles in the obvious, or to use Ahmes' phrase once again, they unravel the world's obscure secrets.

Psychologist Robert Sternberg (1985) described the psychological features of the experience-to-abstraction flow in puzzles in terms of three main processes: (1) the creative use of information that may have originally seemed irrelevant but that has become a source of insight in the solution process; (2) the discovery of a nonobvious relationship between a puzzle and information already stored in memory; and (3) the discovery that nonobvious pieces of information can be combined to form novel ideas. This "triarchic process," as Sternberg calls it, is certainly obvious in many of the puzzles discussed so far.

The term *meme* is, as mentioned above, a term referring to the spread of ideas as if they were genes, put forth by Richard Dawkins (1976). He defines memes as replicating patterns of information (ideas, laws, clothing fashions, artworks, and so on) and of behavior that people inherit directly from their cultural environments. Like genes, memes are passed on with no intentionality on the part of the human organism. Being part of culture, the human organism takes them in unreflectively from birth, passing them on just as unreflectively to subsequent generations. The *memetic code* thus parallels the *genetic code* in directing human evolution. This clever proposal poses an obvious challenge to traditional philosophy, theology, and science. Whether the theory is verifiable or not, it is a useful metaphor, as mentioned, with some modifications, to explain the spread of ideas found in some puzzle format that then become part of mathematics and reside there until activated to produce related ideas. Essentially, a "meme theory" might explain how puzzle archetypes spread. To be clear: the archetype is where the mind's unconscious resources are at work, and these are meaningful; the term *meme* is a communicative notion, indicating how an archetypal idea is passed on culturally.

A well-known example of a puzzle meme spreading to other domains is Euler's Königsberg Bridges Puzzle, as we have seen. One of its memetic descendants is the *Hamiltonian Cycle*, named after the Irish mathematician William Rowan Hamilton, who presented it as a game that he called *Icosian*. The object is to find a cycle along the edges of a dodecahedron such that every vertex is reached a single time, and the ending point is the same as the starting point. The puzzle-game was distributed commercially as a pegboard with holes at the nodes of the dodecahedral graph and was subsequently marketed in Europe in many forms. The solution is a cycle containing twenty (in ancient Greek *icosa*) edges (that is, a Hamiltonian circuit on the dodecahedron) (Figure 5.15)

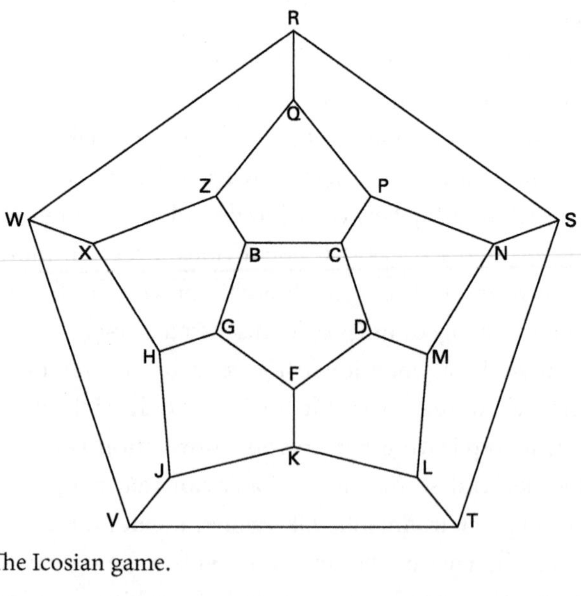

Figure 5.15 The Icosian game.

Hamiltonian circuits must be examined individually, and finding a Eulerian path in them—if one exists—is a matter of trial and error. The surface is a closed one with no break. If we start a path anywhere that cuts the surface at any point, we will end up at the point from which we started. No matter where we penetrate the surface, we are still outside of it. The Hamiltonian cycle has itself become a meme, leading to the famous *Traveling Salesman Problem* (TSP). The TSP is NP-complete, and is often enlisted to show what this question entails, since it is among the most difficult to solve by algorithm. As Elwes (2014: 289) puts it: "If P ≠ NP, then there is some problem in NP which cannot be computed in polynomial time. Being NP-complete, the Travelling Salesman Problem must be at least as difficult as this problem, and so cannot lie in P" (see also Cook 2014).

The TSP has a structure that can be modeled on a computer, given that graphs are computer objects of a specific kind. Here is a standard version of the problem (Benjamin, Chartrand, and Zhang 2015: 122):

> *A salesman wishes to make a round-trip that visits a certain number of cities. He knows the distance between all pairs of cities. If he is to visit each city exactly once, then what is the minimum total distance of such a round trip?*

In this puzzle, all the vertices of a Eulerian graph are to be used. A graph with a Hamiltonian path is thus traceable and connectible. The solution of the TSP is elaborated by Benjamin, Chartrand, and Zhang (2015: 122) as follows (where c = a city, n = number of vertices in a graph). It is worth reproducing it here:

> The Traveling Salesman Problem can be modeled by a weighted graph G whose vertices are the cities and where two vertices u and v are joined by an edge having weight r if the distance between u and v is known and this distance is r. The weight of a cycle C in G is the sum of the weights of the edges of C. To solve this Traveling Salesman Problem, we need to determine the minimum weight of a Hamiltonian cycle in G. Certainly G must contain a Hamiltonian cycle for this problem to have a solution. However, if G is complete (that is, if we know the distance between every pair of cities), then there are many Hamiltonian cycles in G if its order n is large. Since every city must lie on every Hamiltonian cycle of G, we can think of a Hamiltonian cycle starting (and ending) at a city c. It turns out that the remaining (n – 1) cities can follow c on the cycle in any of its (n – 1)! orders. Indeed, if we have one of the (n – 1)! orderings of these (n – 1) cities, then we need to add distances between consecutive cities in the sequence, as well as the distance between c and the last city in the sequence. We then need to compute the minimum of these (n – 1)! sums. Actually, we need only find the minimum of (n – 1)!/2 sums since we would get the same sum if a sequence was

traversed in reverse order. Unfortunately, (n – 1)!/2 grows very, very fast. For example, when n = 10, then (n – 1)!/2 = 181,400.

By translating the physical aspects of the problem (distances, cities, and so on) into symbolic notions, such as paths, weights, and so on that apply to graph systems, we have thus devised a mathematical model of the TSP—a model that decomposes all aspects of the problem into its essential parts. Because of this, it can be translated into an algorithm, as has been done by computer scientists and mathematicians throughout the history of the TSP. The interesting thing here is that it involves knowledge of graph theory and of Hamiltonian cycles—something a computer would not know in advance. But, once programmed, the outputs show many alternatives to the solution—all connected by the main features in the algorithm.

For historical accuracy, it should be mentioned that the TSP was first presented in the 1930s and now constitutes one of the most well-known problems in algorithmic optimization, having led to a large number of programming ideas and methods. As Bruno, Genovese, and Improta (2013: 201) note:

> The first formulation of the TSP was delivered by the Austrian mathematician Karl Menger who around 1930 worked at Vienna and Harvard. Menger originally named the problem the messenger problem and set out the difficulties as follows. At this time, computational complexity theory had not yet been developed: We designate the Messenger Problem (since this problem is encountered by every postal messenger, as well as by many travelers) the task of finding, for a finite number of points whose pairwise distances are known, the shortest path connecting the points. This problem is naturally always solvable by making a finite number of trials. Rules are not known which would reduce the number of trials below the number of permutations of the given points. The rule, that one should first go from the starting point to the point nearest this, etc., does not in general result in the shortest path.

Of course, Menger's challenge has been tackled rather successfully by computer science and mathematics working in tandem today with the development of the field of combinatorial optimization, which was developed to solve problems such as the TSP one. In 1954, an integer programming formulation was developed to solve the problem alongside the so-called "cutting-plane" method, "which enables the finding of an optimal solution (namely, the shortest Hamiltonian tour) for a TSP involving 49 U.S. state capitals" (Bruno, Genovese, and Improta 2013: 202). The problem has been generalized in various ways and studied algorithmically, leading to the growth of optimization theory.

Meme theory is, in the end, not really a theory but a convenient language for explaining how ideas in mathematics are interconnected and how they spread from simple formats to complex ones. The original insights into a puzzle are due to imaginative (archetypal) universals, but their spread is mimetically based.

6

Puzzles and Logic

Logic will get you from A to B. Imagination will take you everywhere.

Albert Einstein (1879–1955)

In the fifth century BCE, philosophical debates broke out throughout Greece over the nature and function of logic in thought. Prominent in them were the philosopher Parmenides (*c.* 510 BCE) and his disciple Zeno of Elea (*c.* 490–430 BCE). The latter became famous for a series of clever arguments that seemed to defy common sense. The arguments came to be known as *paradoxes* (meaning literally "conflicting with expectation"). A group of traveling teachers, called the sophists (from Greek *sophos* "clever"), sided with Zeno, arguing that paradoxes exposed logical thinking as essentially flawed or at least unrealistic. Aristotle dismissed Zeno's paradoxes as exercises in specious reasoning. The central characteristic of the human mind, Aristotle insisted, was its ability to think logically. As clever as they were, Aristotle asserted, paradoxes were ultimately inconsequential because they did not impugn the validity of logic, based on argumentation.

But Aristotle's response was not the end of the matter. On the contrary, the history of logic and mathematics recounts a poetic vindication of Zeno's stand. Moreover, the paradoxes exploit an argumentation logic that Aristotle himself praised. Paradoxes are puzzles in logic. The story goes that one of the most vexing of all the paradoxes concocted during the debates was uttered by Protagoras (*c.* 480–411 BCE), who was the first philosopher to call himself a Sophist. The paradox has come to be known as the Liar Paradox. Its most famous articulation has, however, been attributed to a Cretan named Epimenides in the sixth century BCE. Almost nothing is known about his life, other than the fact that he was a celebrated poet and prophet of Crete. The Liar Paradox continues to stupefy people to this day, since it evokes what we colloquially call "circularity." It has come down to us more or less in the following form:

The Cretan philosopher Epimenides once said: "All Cretans are liars." Did Epimenides speak the truth?

Let's assume that Epimenides spoke the truth. Thus, his statement that "All Cretans are liars" is a true statement. However, from this we must deduce that Epimenides, being a Cretan, is also a liar, making his statement false. But this is a contradiction, since a statement cannot be both true and false. Obviously, we must discard our assumption. Let's assume the opposite, namely that Epimenides is in fact a liar. But, then, if he is a liar, the statement he just made—"All Cretans are liars"—is true. But this is again a contradiction—liars do not make true statements. Obviously, we are confronted with a logical circularity. British logician P. E. B. Jourdain invented an interesting version of the Liar Paradox in 1913, called Jourdain's Card Paradox, which brings out its essential nature in a concrete way (Hsiung 2009):

The following is printed on one side of a card: "The statement on the other side of this card is true." But on the card's other side the statement reads: "The statement on the other side of this card is false." What do you make of the card?

The card makes us go back and forth, from one side to the other, frustratingly. Now, the implications of the Liar Paradox are not inconsequential, as Aristotle and others believed. From antiquity onwards, the goal of logic and mathematics has always been to be free of logical circularity. But it is not. This is why the Liar Paradox has fascinated mathematicians throughout history becoming, over time, one in a series of clever paradoxes that have brought about revolutionary changes in mathematics and philosophy. Despite all the radical implications it has had, ultimately the Liar Paradox does not invalidate the use of practical Aristotelian logic in everyday life. In our three-dimensional world if it is true, for instance, that building A is higher than building B, and that building C is higher than building A, then we can conclude, without any shadow of a doubt, that building C is much higher than building B. Nevertheless, the Liar Paradox continues to warn us against believing that logic is the only path to knowledge. Hunches and experience are probably just as important, if not more so, for grasping the meaning of things, as argued throughout this book.

As Eugene Northrop (1944: 12) aptly puts it, the Liar Paradox is more than a sophism:

The case of the self-contradicting liar is but one of a whole string of logical paradoxes of considerable importance. Invented by the early Greek philosophers, who used them chiefly to confuse their opponents in debate, they have in more recent times served to bring about revolutionary changes in ideas concerning the nature of mathematics.

For some reason, paradoxical statements have a bizarre appeal. They are riddles in logic. They are especially alluring to children, as Henry Dudeney (1958: 15) perceptively observed:

> A child asked, "Can God do everything?" On receiving an affirmative reply, she at once said: "Then can He make a stone so heavy that He can't lift it?"

The child's question is similar to a well-known philosophical conundrum: *What would happen if an irresistible moving body came into contact with an immovable body?* As Dudeney observes, such bizarre paradoxes arise only because we take delight in inventing them. In actual fact, "if there existed such a thing as an immovable body, there could not at the same time exist a moving body that nothing could resist."

Incidentally, it was in sixth-century BCE Greece that the philosopher Heraclitus asserted that the world was governed by the *lógos*, a divine force that produces order in the flux of Nature. *Lógos* came to be viewed shortly thereafter as a rational divine power that directed the universe. Through the faculty of reason, all human beings were thought to share in it. So, in its origins, logic was hardly divorced from mysticism.

This chapter looks at the nature of logic through the filter of logic puzzles and paradoxes, extending the two main notions of previous chapters: namely that (a) puzzles are based on an unconscious archetypal structure and (b) that they become intellectual memes that spread across space and time. The marvelous logic puzzles of Lewis Carroll, Henry E. Dudeney, and others shed ludic light, so to speak, on the nature of so-called *lógos*, showing the relation of logic to the imagination, not its separation from it. In the end, logic can be seen to emerge at the origins of cultures as an organizing grid for the discoveries of the imagination, thus rendering them stable cognitively. All cultures have their own *lógos*, or rules of conduct and understanding that develop from the original mythic dreams and puzzles.

The nature of logic

Many puzzles and games based on chance seem to reveal hidden logical structure in their own unique ways. One of the most famous is the so-called "St. Petersburg Paradox," which takes its name from a paradox proposed by mathematician

Daniel Bernoulli. The paradox was actually devised by his brother Nicolaus Bernoulli (Kraitchik 1942: 138–9). It can be paraphrased as follows:

> *A game played at a casino offers a single player a chance to toss a fair coin in stages. Initially, the stake is 2 dollars and this is doubled each time heads appears. The first time a tail appears, the game ends and the player wins whatever is in the pot. So, a player will win 2 dollars if tails appears on the first toss, 4 dollars if heads appears on the first toss and tails on the second, eight dollars if heads on the first two tosses and tails on the third, and so on. The player thus wins 2k dollars (k = number of tosses). What would be a fair price to pay the casino for entering the game?*

Answering this conundrum involves figuring out the average payout. So, with probability ½, the player wins two dollars; with probability ¼ the player wins four dollars; with probability ⅛ the player wins eight dollars, and so on. The expectation value, E, up to infinity, is however uniform, as shown.

$$
\begin{aligned}
E &= \frac{1}{2}\cdot 1 + \frac{1}{4}\cdot 2 + \frac{1}{8}\cdot 4 + \frac{1}{16}\cdot 8 + \cdots \\
&= \frac{1}{2} + \frac{1}{2} + \frac{1}{2} + \frac{1}{2} + \cdots \\
&= \sum_{k=1}^{\infty} \frac{1}{2} = \infty
\end{aligned}
$$

Proposed solutions to this paradox have led to significant debates within mathematics and to several modifications of probability theory. These need not concern us here. The point is that it brings out the essential characteristics of a form of reasoning that is essential to solving puzzles via logic—we do not play games to maximize expected monetary outcomes, but we act as if we were. Hypothesis thinking is the essence of logical thinking.

When circularity enters into this mental scenario logic breaks down. The source of the circularity in the Liar Paradox, for example, is the fact that it was Epimenides, a Cretan, who made the statement that "All Cretans are liars." It is an example of the problems in logic that arise from *self-reference*, or the situation in which the makers of a statement include themselves in the statement. Bertrand Russell found the paradox to be especially troubling, feeling that it threatened the very foundations of logic and mathematics. To examine the nature of self-reference more precisely, Russell (1918: 228) formulated his own version of the Liar Paradox, called the Barber Paradox:

> *The village barber shaves all and only those villagers who do not shave themselves. So, shall he shave himself?*

The barber is "damned if he does and damned if he doesn't," colloquially speaking. Let's say he decides to shave himself. He would end up being shaved, of course, but the person he would have shaved is himself. And this contravenes the requirement that the village barber shave "all and only those villagers who do not shave themselves." The barber has, in effect, just shaved someone who shaves himself. So, let's assume that the barber decides not to shave himself. But, then, he would end up being an unshaved villager. Again, this goes contrary to the stipulation that he, the barber, shave "all and only those villagers who do not shave themselves"—including himself. It is not possible, therefore, for the barber to decide whether or not to shave himself. Russell argued that such "undecidability" arises because the barber is himself a member of the village. If the barber were from a different village, the paradox would not arise.

To avoid such paradoxes, Russell joined intellectual forces with Alfred North Whitehead in 1913 to outline a set of propositions within a theory of logic that would never produce paradoxes of this kind—that is, like the German philosopher Gottlob Frege (1879), and, of course, Aristotle well before him, Russell sought to find a system of logical argumentation that would exclude self-reference. Using a notion developed two millennia earlier by Chrysippus of Soli, Frege had claimed that circularity could be avoided from logical systems by considering their *form* separately from their *content*. In this way, one could examine the consistency of statements, known more technically as *propositions*, without having them correspond to anything (such as barbers, villages, Cretans, and so on). Frege's approach was developed further by Ludwig Wittgenstein (1922), who used symbols rather than words to ensure that the form of a proposition could be examined in itself for logical consistency separate from any content to which it could be applied. If the statement "It is raining" is represented by the symbol "p" and the statement "It is sunny" by "q", then the proposition "It is either raining or it is sunny" can be assigned the general symbolic form "p v q" (with v = "or"). A proposition in which the quantifier "all" occurs would be shown with an inverted "∀". So, the statement "All Cretans are liars," would be represented as "∀ p". If the form held up to logical scrutiny, then that was the end of the matter. The problem, Wittgenstein affirmed, was that we expect logic to interpret reality for us. But that is expecting way too much from it. Wittgenstein's system came to be known as "symbolic logic"—a system of representation prefigured by Lewis Carroll in his ingenious book *The Game of Logic* (reprinted 1958a).

Russell introduced the theory of "types," whereby certain types of propositions only would be classified into different levels (more and more abstract) and thus considered separately from other types. This seemed to avoid the problem of self-reference—for a while anyhow. The Polish mathematician Alfred Tarski (1933) developed Russell's idea further by naming each level of increasingly abstract statements a *metalanguage*. A metalanguage, Tarski showed, is essentially a statement about another statement. At the bottom of the hierarchy are straightforward statements about things such as: "Earth has one moon." Now, if we say, "The statement that Earth has one moon is true," we are using a different type of language, because it constitutes a statement about a previous statement. It is a metalanguage. The problem with this whole approach is, of course, that more and more abstract metalanguages are needed to evaluate lower-level statements. And this can go on ad infinitum. In effect, Tarski's system only postpones making final decisions about "what is what."

The problem may be that mathematicians have confused logic (the assumed metalanguage or metastructure of mathematics) with mathematics. As Charles Peirce argued eloquently, the two are ontologically different. This is what he wrote *c.* 1906 (in Kiryushchenko 2012: 69):

> The distinction between the two conflicting aims [of logic and mathematics] results from this, that the mathematical demonstrator seeks nothing but the solution of his problem; and, of course, desires to reach that goal in the smallest possible number of steps; while what the logician wishes to ascertain is what are the distinctly different elementary steps into which every necessary reasoning can be broken up. In short, the mathematician wants a pair of seven-league boots, so as to get over the ground as expeditiously as possible. The logician has no purpose of getting over the ground: he regards an offered demonstration as a bridge over a canyon, and himself as the inspector who must narrowly examine every element of the truss because the whole is in danger unless every tie and every strut is not only correct in theory, but also flawless in execution. But hold! Where am I going? Metaphors are treacherous—far more so than bridges.

Metalanguages are fraught with logical loopholes, as the Russell-Whitehead-Tarski episode showed. The origin of the debate on logic as a metalanguage, actually, goes back to Euclid and Aristotle, although they did not use the term metalanguage (of course). They developed the method of proof as a logical way of demonstrating some idea or concept as being valid or invalid through reasoning. The validity of an argument depended on the logical consistency of the argument, not on the truth or falsity of its premises. This view of logic was mirrored in what Aristotle called the categorical syllogism, which is based on connecting premises, such as the following, to each other:

1. All mammals are warm-blooded.
2. All cats are mammals.
3. Therefore, all cats are warm-blooded.

This syllogism is valid because the premises are connected logically. Each is composed of *categorical* terms (terms that denote categories such as mammals, cats, and so on). Each of the premises has one term in common with the conclusion: the major term (1) which forms the predicate of the conclusion (3), and the minor term (2), which forms the subject of the conclusion (3). The categorical term in common in the premises is called the middle term. Above it is *mammals*. The skeletal structure of the categorical logic of the above syllogism can be shown as follows:

1. All A are B.
2. All C are A.
3. Therefore, all C are B.

One does not need to use syllogistic reasoning to accept this as true, though. Common sense tells us that this is so. However, common sense does not show us the validity of the logic behind the argumentation. Logical proof brings about a sense of certainty about observed phenomena. One could measure the angles in a triangle ad infinitum and they would always add up to 180°, with some minor variation for human error in the measuring process. But when it was demonstrated that this is so by logical proof then the observation became a proposition that needed no further exploration, thus eliminating the need to keep on measuring triangles.

It was George Boole (1854) who used the idea of sets to unite logic and mathematics into a general method of analysis. To test an argument, Boole converted statements into symbols, in order to show their logical structure, independently of their meanings. Then, through rules of derivation, he showed that it is possible to determine what new propositions may be inferred from the original premises. Boole's primary objective was to break down the logic of proof into its bare abstract structure by replacing words and sentences (which bear contextual or categorical meaning) with symbols (which do not). This, he claimed, would enucleate the essence of logical proof. Moreover, he reduced symbolism to its oppositional structure—using the one of the binary system to stand for *true* and the zero for *false*. Instead of addition, multiplication and the other operations of arithmetic (which bear historical meanings) he used the conjunction (∧), disjunction (∨), and complement or negation (¬), in order to divest operations from any kind of external-contextualized information they may bring to bear on the logic used. These operations can be expressed either with truth tables or Venn diagrams, which show how sets *x* and *y*, relate to each other logically.

The Boolean system gave a concrete slant to the question of what is mathematics and what its relation to logic is. Moreover, it forced mathematicians to reconsider their definitions, axioms, and assumptions from the perspective of logical entailment, taking nothing for granted. This was Giuseppe Peano's aim in 1889 (Peano 1973), who wanted to formalize the operations of arithmetic by breaking them down into their elemental logical components, recalling Euclid's axioms for geometry. His nine axioms start by establishing the first natural number (no matter what numeral system is used to represent it), which is zero. The other axioms are *successor* ones, showing that they apply to every successive natural number. For example, the second axiom states that for every natural number x, x = x; the third one, then, follows with the statement that, for all natural numbers x and y, if x = y then y = x; the fourth then states that, for all natural numbers x, y and z, if x = y and y = z, then x = z; and so on.

Peano's axioms may seem like self-evident concepts, but the goal of all logical systems is to make the obvious, obvious. If one were to program a machine to carry out arithmetical operations, it would need to have these axioms built into its appropriate algorithmic structure. They cannot be taken for granted as "obvious." On the other hand, while the axioms are of course valid logically, they do not tell the whole story of mathematics and especially of how we come to devise such systems in the first place. They are useful, not revelatory. As Stewart (2013: 313) observes, the use of the term *exist* in any logical treatment of mathematics raises several issues, the most obvious one being the definition of *exist* itself:

> The deep question here is the meaning of 'exist' in mathematics. In the real world, something exists if you can observe it, or, failing that, infer its necessary presence from things that can be observed. We know that gravity exists because we can observe its effects, even though no one can see gravity. However, the number two is not like that. It is not a thing, but a conceptual construct.

The irrational numbers and the imaginary ones did not "exist" until they cropped up in the solution of two specific equations made possible by the Pythagorean theorem and the concept of quadratic equation respectively. So, where were they before? Were they waiting to be discovered? This question is clearly at the core of the nature of discovery in mathematics. The same story can be told over and over within the field—transfinite numbers, graph theory, and so on. These did not "exist" until they crystallized in the conduct of mathematics, through ingenious notational modifications, clever insights, ludic explorations with mathematical forms, and so on.

Russell was well aware of the inbuilt pitfalls of axiomatic logic and its connection to natural language, asking the question of whether any part of mathematics can actually be proven (Russell and Whitehead 1913), and, if so, what kind of metalanguage, other than words and statements, could be used. How, for example, can we prove that 1 + 1 = 2, even if we articulate this to be a logical derivation from previous axioms? Russell was determined to put mathematics on a solid logical footing. Using connective symbols, Russell did, indeed, prove that 1 + 1 = 2, in a way that at first seemed to be non-tautological. But, shortly thereafter, Kurt Gödel showed in 1931 that any formal theory of basic mathematical truths and their provability is inconsistent if it includes a statement about its own consistency. In other words, when mathematicians attempt to lay a logical basis to their craft, or try to show that logic and mathematics are one and the same, they are playing a mind game that is bound to come to a halt, as Alan Turing (1936) also argued a few years later. Turing asked if there is a general procedure for deciding if a self-contained computer program will eventually come to a halt. One cannot decide if the program will stop when it runs with a given input. Turing started with the assumption that the halting problem was decidable and then constructed an algorithm that halts if and only if it does not halt, which is a contradiction. In their 1986 book, *The Liar*, mathematician Jon Barwise and philosopher John Etchemendy adopted a practical view of the pitfalls of self reference. They argued that it arises only because we allow it to arise. The meaning of a statement can only be determined by assessing the context in which it is uttered along with our reasons for constructing it. Once such factors are determined, no self-reference or undecidability paradoxes arise.

As just mentioned, the problem of undecidability was tackled ingeniously in 1931 by German logician Kurt Gödel, who showed why it is a fact of human life, no matter how hard we try to eliminate it from our logical systems. In models like the Russell-Whitehead one, it was taken for granted that every proposition within it could be either proved or disproved within the system of propositions. But Gödel startled the mathematical world by showing that this was *not* the case. He showed that a logical system invariably contains a proposition that is true but unprovable within it. For illustrative purposes, Gödel's ingenious demonstration can be paraphrased as follows:

> Consider a mathematical system that is both *correct*—in the sense that no false sentence is provable in it—and contains a sentence *S* that asserts its own unprovability in the system. *S* can be formulated simply as: "I am not provable in system *T*." What is the truth status of the sentence *S*? If it is false, then its opposite is true, which means that *S* is provable in system *T*, contrary to our assumption that no false sentence is provable in the system. Therefore, we conclude that *S* must be true, from which it follows that *S* is unprovable in *T*, as *S* asserts. Thus, *S* is true, but not provable in the system.

Gödel proved, in effect, that any consistent (correct) logical system is incomplete, because formulae (sentences, propositions, and so on) can be constructed that can neither be proved nor disproved within the system. His demonstration thus shattered the dream of building a logical edifice of consistent propositions, resting on the cornerstone of Aristotelian logic. The late American logician Raymond Smullyan (1997: 152) provides a clever puzzle version of Gödel's proof as follows:

> *Let us define a logician to be accurate if everything he can prove is true; he never proves anything false. One day, an accurate logician visited the Island of Knights and Knaves, in which each inhabitant is either a knight or a knave, and knights make only true statements and knaves make only false ones. The logician met a native who made a statement from which it follows that the native must be a knight, but the logician can never prove that he is! What statement would work?*

The statement that would work is: *You cannot prove that I am a knight*. If the native were a mendacious knave, then his statement would be false. If the native is actually a knave and the logician would be able to prove it, then we would contravene the given condition that the logician is accurate and thus incapable of proving anything false. Therefore, the native must be a knight. This means that his statement is true. But if his statement is true, then the logician cannot prove that the native is a knight—the statement declares as much. So, even though the native is a knight, the logician will never be able to prove it.

The Monty Hall problem

Logic is often associated with "common sense," as discussed above. But the unreliability of this association is the source of various kinds of puzzles. One of these actually arose from a situation connected to a television quiz show, and was called the "Monty Hall, Problem" (MHP), after television quiz show host Monty Hall, who was the presenter of *Let's Make a Deal*, which premiered in 1963 and continued until the 2010s on American television. It was formulated as a puzzle in probability theory by Steven Selvin in 1975, showing that common-sense logic is subject to the laws of chance, not impervious to them. The contestants on the show had to select from three doors that hid different prizes. The situation can be broken down into stages as follows:

1. A contestant has to choose one of three doors: *A*, *B*, and *C*. Behind one is a new car, behind the other two are goats.
2. The contestant chooses one door, say *A*.
3. He or she has a ⅓ probability of selecting the car.
4. Monty Hall knows where the car is, so he says: "I'm not going to tell you what's behind door *A*, yet. But I will reveal that there is a goat behind door *B*."
5. Then he asks: "Will you now keep door *A* or swap to *C*?"

The assumption of most people is that the odds are 50/50 between *A* and *C*, so that switching would make no difference. But that is incorrect: *C* has a ⅔ probability of concealing the car, while *A* has just a ⅓ probability. This seems to defy common sense, but probability reasoning says something different. Elwes (2014: 334) explains it as follows:

> It may help to increase the number of doors, say, to 100. Suppose the contestant chooses door 54, with a 1% probability of finding the car. Monty then reveals that doors 1–53, 55–86, and 88–100 all contain wooden spoons. Should the contestant swap to 87, or stick with 54? The key point is that the probability that door 54 contains the car remains 1%, as Monty was careful not to reveal any information which affects this. The remaining 99%, instead of being dispersed around all the other doors, become concentrated at door 87. So she should certainly swap. The Monty Hall problem hinges on a subtlety. It is critical that Monty knows where the car is. If he doesn't, and opens one of the other doors at random (risking revealing the car but in fact finding a wooden spoon), then the probability has indeed shifted to ½. But in the original problem, he opens whichever of the two remaining doors he knows to contain a wooden spoon. And the contestant's initial probability of ⅓ is unaffected.

For the sake of historical accuracy, it should be mentioned that the MHP is similar to the "Three Prisoners Problem" devised by Martin Gardner in 1959 (see Gardner 1961)—suggesting that it is a puzzle meme of sorts. We need not discuss Gardner's version here since it reveals the exact same kind of reasoning. Of course, playing by the rules of probability means nothing if one loses—that is, finding oneself on a wrong point in the probability curve—and end up losing the car. However, knowing about the existence of the curve leads to many more insights into the nature of real events than common sense. The MHP has had various implications that reveal the power of probability reasoning to unravel the hidden structure of real-life events. Our assumption that "two choices means 50–50 chances" is true when we know nothing about either choice. If we picked any coin, then the chances of getting a head or tail are, of course, 50–50. But if we are told that one of the sides will come up more often because it has been somehow designed physically to do so, then everything changes. Indeed, information is what matters here and changes the outcome.

So, the MHP brings out the principle that the more we know, the better our decision will be. If the number of doors in the MHP were 100 this becomes even clearer, as Elwes demonstrated. As Monty starts eliminating the bad candidates (in the ninety-nine that were not chosen), he shifts the focus away from the bad doors to the good ones more and more. After Monty's filtering, we are left with the original door and the other door. Here is where probability reasoning comes into play, allowing us to generalize the MHP—the probability of choosing the desired door improves as we get more information. Without any evidence, two choices are equally likely. As we gather additional evidence (and run more trials) we can increase our confidence interval that A or C is correct:

1. Two choices are 50–50 when we know nothing about them.
2. Monty intervenes by "filtering" the bad choices on the other side.
3. In general, the more information, the more the possibility of re-evaluating our choices.

Like the St. Petersburg puzzle, the MHP shows that decision-making is implanted in probability reasoning, not common-sense logic. Overcoming the instinctive reaction that "two choices means 50–50 chances" is the biggest hurdle in grasping the MHP. The more we know the better our decision; the computation of probabilities provides the basis on which to make advantageous decisions.

Paradoxes and cognition

Paradoxes are products of *homo ludens*—they are logical legerdemain, showing how logic can be turned upside down to show its fallibility. But paradoxes, like most puzzles, are also sources of discovery. For example, the calculus is one of the most important branches of mathematics, without which it would be impossible to conduct research in physics and other sciences, not to mention carrying out engineering projects. Arguably, it can be traced to a set of memes—the paradoxes of Zeno of Elea. Zeno tried cleverly to show that motion, change, and plurality (reality consisting of many substances) are impossible. He used *reductio ad absurdum* (proof by contradiction) reasoning, deriving seemingly impossible conclusions from logical premises. Zeno actually devised at least forty such paradoxes, but only eight have survived. These were called "dialectical" by Aristotle, who tried to dismiss them, but essentially could not. Zeno's paradoxes concerning motion make up his most famous surviving ones. In one, Zeno argued that if we use logic, then we must conclude that a runner will never reach the end of a race course, even though he will actually do so if we just look at him run. He argued this by stating that the runner first completes half of the course, then half of the remaining distance, and so on infinitely, without ever reaching the end. The successive stages of the runner's location along the race course form an infinite series with each term in it half of the previous one:

$$\{1/2 + 1/4 + 1/8 + 1/16 + 1/32 + \ldots\} = 1/2^n$$

Zeno argued that the runner would never cross the end line, even though we know by experience that he actually does. Zeno's startling argument, which could not be dismissed logically, raised profound issues about time, space, and infinity. In it, we can easily discern the concept of *limits* which, in turn, inspired the invention of the calculus. Zeno's paradoxes treat distance and time as if they can be segmented into infinitely smaller parts or points. Let's look more closely at the reasoning involved in the above paradox, since it has many implications for deconstructing logical thinking. The paradox states that a runner would never be able to cross a finish line, if one used logical reasoning.

The runner must first traverse *half* the distance to the finish line. Then, from mid-position, the runner faces a new, but similar, task—he must traverse *half of the remaining distance* between himself and the finish line. But from that new position, the runner then faces the same task—he must once more cover *half of the new remaining distance* between himself and the finish line. Although the successive half distances between himself and the finish line would become increasingly (indeed, infinitesimally) small, the crafty Zeno concluded that the runner would come very close to the finish line, but would never cross it. So, the logical reasoning used seems to betray us, given that in truth the runner does cross the line, unless something untoward happens to him along the way. As mentioned, the paradox cannot be easily dismissed as specious argumentation. Indeed, in it we can see the archetype for the notion of infinite series. In later mathematics, the successive distances that the runner must cover are said to form a geometric series, each term of which is half the one before. The sum of the terms in the series will never reach "1," the whole distance to be covered. So, "1" is defined as a limit. This constitutes a veritable revolution in thought. It upsets our notions of "common sense," and leads to unraveling an "obscure secret" about time and space, to use Ahmes' rather appropriate expression once again.

That unraveling came from Sir Isaac Newton and Gottfried Wilhelm Leibniz, who were likely influenced by the memes emanating from Zeno's paradoxes, coming up, independently, with an ingenious, yet remarkably simple, solution to them. They simply asserted that the sum to which a series such as {1/2, 1/4, 1/8, 1/16, . . .} converges as it approaches infinity is the distance between the starting line and the finish line. Thus, the limit of the runner's movement is, in fact, the unit distance of "1". It is beyond the scope of the present treatment to discuss the historical details of the origins of the calculus. Suffice it to say that it was a radical paradigm shift in mathematical thinking. Indeed, when the calculus was first proposed it met with acerbic criticism from philosophers and skepticism by some mathematicians. But the new branch of mathematics easily survived such attacks, for the simple reason that it provided a powerful new conceptual framework for answering the classical unsolved problems of physics and the paradoxes of logic.

Perhaps the central topic in the study of logic is the following one: Is logic related to language? The answer by Lakoff and Núñez (2000) is that the two share the same neuro-conceptual substrate. Their basic claim is that the proofs and theorems of mathematics are arrived at, initially, through the same cognitive mechanisms that underlie language—analogy, metaphor, and metonymy, a claim that has been substantiated with neurological techniques such as fMRI and other scanning devices (Danesi 2016). Finding where this substrate is located and determining its characteristics is an ongoing goal of neuroscience.

Gödel's (1931) famous proof, as Lakoff has argued, was itself a metaphorical blend, inspired by Cantor's diagonal method, which need not concern us here. As mentioned, Gödel proved that within any formal logical system there are results that can be neither proved nor disproved. Gödel found a statement in a set of statements that could be extracted by going through them in a diagonal fashion—now called Gödel's diagonal lemma. That produced a statement, S, like Cantor's C, that does not exist in the set of statements. Cantor's diagonal and one-to-one matching proofs are metaphorical blends—associations linking different domains in a specific way (one-to-one correspondences). This insight led Gödel to envision three metaphors of his own: (1) the "Gödel number of a symbol," which is evident in the argument that a symbol in a system is the corresponding number in the Cantorian one-to-one matching system (whereby any two sets of symbols can be put into a one-to-one relation); (2) the "Gödel number of a symbol in a sequence," which is manifest in the demonstration that the nth symbol in a sequence is the nth prime raised to the power of the Gödel number of the symbol; and (3) "Gödel's central metaphor," which is Gödel's proof that a symbol sequence is the product of the Gödel numbers of the symbols in the sequence.

The proof exemplifies how neurological blending works. When the brain identifies two distinct entities in different neural regions as the same entity in a third neural region, they are blended together. Gödel's metaphors come from neural circuits linking a *number* source to a *symbol* target. In each case, there is a blend, with a single entity composed of both a *number* and a *symbol sequence*. When the symbol sequence is a formal proof, a new mathematical entity appears—a "proof number." But the whole process was triggered by a meme—Cantor's diagonal method—which traces its source to the human imagination and its ability to think about the world in real terms by connecting parts within the world in a holistic way.

René Thom (1975) referred to discoveries in mathematics as "catastrophes" in the sense of events that subvert or overturn existing knowledge. Thom named the process of discovery as "semiogenesis" or the emergence of "pregnant" (suggestive) forms within symbol systems themselves. These emerge by happenstance through contemplation and manipulation of the systems. As this goes on, every once in a while, a catastrophe occurs that leads to new insights, disrupting the previous system. Now, while this provides a description of what happens—discovery is, indeed, catastrophic—it does not tell us why the brain produces catastrophes in the first place. Perhaps the connection between the brain, the body, and the world will always remain a mystery, since the brain cannot really study itself. This connection, as discussed throughout this book, is embedded in many puzzles.

It is no exaggeration to claim that science works like puzzles—it is based on guesses, models, and probable outcomes. To make their hunches useable or practicable, scientists express them in mathematical language, which gives them a shape that can be seen, modified, and tested. Physicist Lee Smolin (2013) asks a simple but profound question related to the foregoing discussion: Can the laws of science account for the highly improbable set of conditions that triggered the Big Bang jump-starting the universe? The question Smolin (2013: 46) asks is, essentially, whether or not the mathematics is correct, but the science is not:

> Logic and mathematics capture aspects of nature, but never the whole of nature. There are aspects of the real universe that will never be representable in mathematics. One of them is that in the real world it is always some particular moment.

As a previous model breaks down under the weight of new facts, it is discarded and replaced with a new one. This process was called *falsification* by philosopher Karl Popper (1935, 1963). As he put it (1963: 34): "Every genuine *test* of a theory is an attempt to falsify it, or to refute it. Testability is falsifiability; but there are degrees of testability: some theories are more testable, more exposed to refutation, than others; they take, as it were, greater risks." In other words, it only takes one counter-observation to falsify an existing theoretical paradigm. Science progresses when a theory is shown to be falsifiable and a new theory is introduced which better explains the observed phenomena. That is what happened with quantum physics. A theory is scientific, then, if we can show what would possibly cause us to reject it.

It was Thomas Kuhn (1970) who coined the term "paradigm shift" to describe how progress in science occurs. Kuhn attacks the "development-by-accumulation" view of science, which holds that science progresses linearly by accumulation of theory-independent facts. In this paradigm, older theories give way successively to wider, more inclusive ones. Like Popper, he agrees that scientists have a worldview or paradigm that they bring to their observations. Paradigms shift all the time on the basis of new ideas. Scientists accept the dominant paradigm until anomalies appear. Then, they begin to question the basis of the paradigm, with new theories which challenge it. Eventually, one of the new theories becomes accepted as the new paradigm. An argument made here is that the paradigm is often changed via a puzzle meme that may shatter the previous one with the insights it bears.

Logic deconstructed

Puzzles and paradoxes in logic ultimately allow us to deconstruct the nature of logic in its various modalities. One of the most revelatory in this respect was invented by British puzzle-maker Hubert Phillips (1933, 1934, 1966). The following is a version of one of his original puzzles:

> *The people of an island culture belong to two tribes—the Bungu and the Mungu. Since members of both tribes look and dress alike, and since they speak the same*

language, they are virtually indistinguishable. It is known, however, that the members of the Bungu tribe always tell the truth, whereas the members of the Mungu tribe always lie. The anthropologist who became interested in their fascinating social system, Dr. Mary Truthspin, recently came across three men on the island.

"To which tribe do you belong?" Dr. Truthspin asked the first.

"Dutu luneh," he replied in his native language.

"What did he say?" she asked the second and third men, both of whom had learned to speak some English.

"He said that he is a Bungu," said the second.

"No, he said that he is a Mungu," said the third.

Can you figure out to which tribes the second and third men belonged?

The method used in solving this genre of puzzle is to zero in on a specific statement in order to test its truth or consistency, or else, as in this case, to unravel its actual content. The key, therefore, lies in translating *Dutu luneh* into English. Assume that the first individual belonged to the truth-telling Bungu tribe. His answer in English to the anthropologist's question *To which tribe do you belong?* would have been, of course, that he belonged to the Bungu tribe, because as a Bungu he would not lie. So, in this hypothetical scenario, *Dutu luneh* translates as *I belong to the Bungu tribe.* Now, assume the opposite, namely that the first individual belonged to the mendacious Mungu tribe. His answer to Truthspin's question would have been a lie. He certainly would not have admitted to belonging to the Mungu tribe. Instead, he would have lied, and said that he belonged to the other tribe, the Bungu. Once again, in this second hypothetical scenario, *Dutu luneh* translates as *I belong to the Bungu tribe*. In sum, no matter to which tribe the first man really belonged, the anthropologist would have gotten the same answer from him. Now, consider the responses given by the other two men to Truthspin's follow-up question—*What did he say?* We start with the second individual's response:

Dr. Truthspin:	What did he say?
Second Man:	He said that he is a Bungu.

As we have just discovered, the first man did, indeed, say that he was a Bungu. So, the second told the truth. That means that he himself was a member of the Bungu tribe. Finally, consider the response given by the third man:

Dr. Truthspin:	What did he say?
Third Man:	He said that he is a Mungu.

As we know, the first man said that he was a Bungu. So, the third clearly lied. This means, of course, that he was a member of the deceitful Mungu tribe. It is, of course, not possible to determine to which tribe the first man belonged. In sum, this type of puzzle really zeroes in on how we must keep inferences free of assumptions and decode statements in terms of their meanings, rather than as part of propositions. Phillips' puzzle is, in effect, a lesson on the weakness of common-sense logic, and how we must constantly make inferences.

In the 1970s, Raymond Smullyan became the undisputed master of this type of logic puzzle. Here is an example of his delightful puzzle art (Smullyan 1997: 48–9):

> *"News has reached me, O Auspicious King, of a curious town in which every inhabitant is either a Mino or an Amino."*
>
> *"Oh my goodness, what are they?" asked the king.*
>
> *"The Minos are worshippers of a good god; whereas the Aminos worship an evil god. The Minos always tell the truth—they never lie. The Aminos never tell the truth—they always lie. All members of one family are of the same type. Thus given any pair of brothers, they are either both Minos or both Aminos. Now, I heard a story of two brothers, Bahman and Perviz, who were once asked if they were married. They gave the following replies:*
>
> *Bahman: We are both married.*
>
> *Perviz: I am not married.*
>
> *Is Bahman married or not? And what about Perviz?*

We know that the two statements are either both true or both false, since the brothers are of the same religion. They obviously cannot both be true, since they contradict each other. So, they are both false. Because Perviz's statement is false, we can take from it the opposite of what he said: Perviz is married. Bahman's statement is also false; they are not both married. Since we have established that Perviz is married, we conclude that Bahman is not. Smullyan follows this puzzle up with a more challenging one:

> *"According to another version of the story, O Auspicious King, Bahman didn't say that they were both married; instead he said, 'We are both married or both unmarried'." If that version is correct, then what can be deduced about Bahman and what can be deduced about Perviz?*

It is useful to write out the new pair of statements, for the sake of clarity:

Bahman: We are both married or both unmarried.

Perviz: I am not married.

If both statements are true, then both brothers are unmarried. If both statements are false, then Perviz is married, contrary to what he says. Now, since

Bahman's statement is false, and Perviz is married, Bahman is unmarried. In either case, we can conclude that this is true.

Phillips characterized the intellectual pleasure that such puzzles hold as an *aesthetics of mind*. He put it as follows (1934: iv): "The invention of such exercises, and the solving of them, both give great pleasure, since their construction can involve—and in my view should have reference to—principles of artistry which embody an aesthetic of their own." In fact, Phillips's characterization applies to the pleasure, or Aha Effect, that ensues from solving virtually any kind of puzzle. A puzzle is, indeed, a small "work of art" that stimulates curiosity and provides a kind of aesthetic pleasure all its own. While any puzzle can produce this effect, a cursory examination of the classic puzzles that have withstood the test of time indicates that the main aesthetic-producing features are the unexpectedness of the answer and the simplicity of the solution when revealed.

The aesthetic index, as it can be called (Danesi 2002), is also evident in a type of puzzle that simulates the kind of reasoning used by, say, detectives in solving a case. Dudeney's (1958) brilliant deduction puzzles are cases in point. Here is a paraphrase of his original puzzle:

> *In a certain company, Bob, Janet, and Shirley hold the positions of director, engineer, and accountant, but not necessarily in that order. The accountant, who is an only child, earns the least. Shirley, who is married to Bob's brother, earns more than the engineer. What position does each person fill?*

This genre of puzzle truly brings out what logical reasoning is all about. We are told that: (1) the accountant is an only child, and (2) Bob has a brother (to whom, incidentally, Shirley is married). So, what can we deduce from these two facts? Clearly, we can eliminate Bob as the accountant—an only child. We are also told that the accountant earns the least of the three, and that Shirley earns more than the engineer. From these two facts, two obvious things about Shirley can be established: (1) she is not the accountant (who earns the least, while she earns more than someone else); (2) she is not the engineer (for she earns more than the engineer). So, she is the director. By elimination, we deduce that Janet is the accountant. Also by elimination, we deduce that Bob is the engineer.

As Dudeney ingeniously showed with this genre of logic puzzle, logical reasoning consists of several cognitive processes working in tandem, such as elimination, deduction, and deriving unique conclusions from specific facts and associations of the facts. This brings about not so much an Aha Effect as a sense of aesthetic satisfaction that comes from deducing facts from given information by connecting them logically. As Phillips (1937: vii) put it, solving puzzles like the one above produces an intellectual "kick," which results from discovering the patterns they conceal. The aesthetic index seems to be inversely proportional to the complexity of its solution or to the obviousness of the pattern, trap, or trick it hides. Simply put, the longer and more complicated the answer to a puzzle, or the more obvious it is, the less appealing the puzzle seems to be. Puzzles with simple, yet elegant solutions, or puzzles that hide a nonobvious principle, have a higher aesthetic index. As John Allen Paulos (1991: 113) argues, puzzles are forms of intellectual play, based on the same mental plan on which humor is rooted. In both cases, getting to the "punch line" is the source of the pleasure. The less obvious the punch line, the funnier the joke; the less obvious the answer, the more pleasurable the puzzle.

Deductive reasoning emerged in Greece in the 500s BCE with Thales, the founder of Greek philosophy and one of the "Seven Wise Men of Greece." His ingenious idea caught on broadly and was adopted by practitioners of arithmetic and geometry instantly, including Pythagoras, becoming the epistemological foundation for the unification of the two crafts into one discipline. The unification was elaborated formally in Euclid's *Elements*, where he established general principles, called theorems, from specific cases through various methods of proofs—all of them based on logical reasoning.

Historically, the methods of proof have always been adaptive and rather ingenious. As an example, let's revisit Euclid's proof that prime numbers are infinite—one of the first uses of *reductio ad absurdum*, the method introduced by Zeno (as we have seen). The integers are divided into numbers that can be decomposed into factors—composite numbers—and those that cannot—prime numbers. The numbers 12, 42, and 169, for instance, are all composite because they are the products of prime factors: $12 = 2 \times 2 \times 3$, $42 = 7 \times 2 \times 3$, $169 = 13 \times 13$. The first primes are: 2, 3, 5, 7, 11, 13, 17, 19, 23. Now, even a cursory examination of the number line—{0, 1, 2, 3, 4, 5, 6, 7, 8, 9, 10, . . .}—reveals that there are fewer and fewer primes as the numbers increase. Thus, it appears logical to conclude that the primes must come to an end at some point. But, with a blend of Aha thinking and logical reasoning, Euclid proved that this is not so.

He started with the assumption that there may, indeed, be a finite set of primes, labeling them as follows: $\{p_1, p_2, p_3, \ldots p_n\}$. This is the basic step in *reductio ad absurdum* logic—a method of proving the falsity of a proposition by showing that its logical consequence is absurd or contradictory. The symbol p_n stands for the last (largest) prime. The set would look like this: $\{2, 3, 5, 7, \ldots p_n\}$. Euclid then obviously had an Aha insight: What kind of number would result from multiplying all the primes in the set: $\{p_1 \times p_2 \times p_3 \times \ldots \times p_n\}$? The result would, of course, be a composite number because it can be factored into smaller prime factors—p_1, p_2, and so on. Then Euclid added 1 to this product: $\{p_1 \times p_2 \times p_3 \times \ldots \times p_n\} + 1$. Now, this number is not decomposable, because when any of the prime factors available to us $\{p_1, p_2, p_3, \ldots p_n\}$ are divided into it, a remainder of 1 would always be left over. So, the number $\{p_1 \times p_2 \times p_3 \times \ldots p_n\} + 1$ is either: a prime number that is obviously much greater than p_n; or a composite number with a prime factor that, as just argued, cannot be found in the set $\{p_1, p_2, p_3, \ldots p_n\}$ and is thus also greater than p_n. Either way, there must always be a prime number greater than p_n. Euclid showed with his ingenious logic that the primes never end.

Sudoku

Perhaps no other contemporary recreational puzzle genre exemplifies the use of logical reasoning more than does Sudoku. Like Dudeney's puzzle above, it involves trial and error, deduction and an occasional Aha insight. It is such a well-known pastime that it needs very little explanation here. The basic puzzle is made up of a nine-by-nine grid, divided into three-by-three subgrids. The diagram shows some numbers from one to nine that have been already inserted into their appropriate cells. The goal of the puzzle is to fill in the remaining empty cells, one number in each, so that each column, row, and box contains all the numbers from one to nine, and so that no number is repeated in any row, column or inside a box (Figure 6.1).

1	9	7	4	2	3		6	5
	4			6				9
	5		8		7	4	2	1
7	1		5	3	8	2		
	3	6		1			9	8
8				4		7	1	3
9		2	3		4	1	5	7
3	7		2			9		4
5	8	4	9		1		3	2

Figure 6.1 Typical Sudoku puzzle.

Looking at the empty cell in the top row of the grid, we note that it is missing the number 8. Also the number 6 is missing from the right-most column. We put both digits in these cells. Now, looking at the row starting with 7 (the fourth row from the top), we note that there are two empty cells, which must contain 4 and 9 in some order, since they are the two missing digits in the row. If we put the 9 in the second-last cell in the row (Figure 6.1), we will produce a repetition of the 9 in the column that crosses the row. So, we must put the 4 there. Using the same type of reasoning, we continue inserting numbers in the cells until all digits are inserted according to the two rules.

1	9	7	4	2	3	8	6	5
2	4	8	1	6	5	3	7	9
6	5	3	8	9	7	4	2	1
7	1	9	5	3	8	2	4	6
4	3	6	7	1	2	5	9	8
8	2	5	6	4	9	7	1	3
9	6	2	3	8	4	1	5	7
3	7	1	2	5	6	9	8	4
5	8	4	9	7	1	6	3	2

Figure 6.2 Solution to the Sudoku puzzle.

There are many variations to the puzzle's format, with such colorful names as Killer Sudoku and Samurai Sudoku. Killer Sudoku has a more complicated objective than the basic puzzle. It introduces the *cage*, which is a different group of cells denoted by different coloring or other signs. The puzzle must be completed based on the following rules: Each region must contain only one number; the sum of all the numbers in one cage must be identical to the number in the cage; and no number must appear twice or more in a cage. Samurai Sudoku consists of five overlapping Sudoku grids in the corners. Sequential grids also exist, with values in specific locations required to be transferred to others.

As Nuessel (2013) elaborates, KenKen is the most mathematically useful derivative of the Sudoku concept. The first such puzzles appeared in *The Times of*

London and then in *The New York Times*. The instructions for doing a KenKen puzzle are as follows: The grid must be filled with digits that do not repeat in any row or column; and the digits in each heavily outlined box must produce the target number shown, by using addition, subtraction, multiplication or division, as indicated in the box. A four-by-four grid uses the digits 1, 2, 3, 4, a six by six grid will use the digits 1, 2, 3, 4, 5, 6. The puzzle was invented by a Japanese mathematics instructor, Tetsuya Miyamoto, to teach his students arithmetic, logic, and patience.

Logic and imagination

The puzzle art of Hubert Phillips (1933, 1934, 1966) actually exemplifies how logic and imagination form a blended system of thought. Here is another of Phillips' classic puzzles:

> *Before they are blindfolded, three women are told that each one will have either a red or a blue cross painted on her forehead. When the blindfolds are removed, each is supposed to raise her hand if she sees a red cross and to drop her hand when she figures out the color of her own cross. Now, here's what actually happens. The three women are blindfolded and a red cross is drawn on each of their foreheads. The blindfolds are removed. After looking at each other, the three women raise their hands simultaneously. After a short time, one of the women, lowers her hand and says, "My cross is red." How did she figure it out?*

Solving this puzzle hinges on projecting oneself "into the minds" of the characters of the puzzle, so as to envision what each one thinks as she looks at the blindfolds. We can start, for the sake of clarity, by calling the three women A, B, and C. Let's assume that A is the one who figured out the color of the cross on her head. Let us now deconstruct A's thinking process. She looks at B and C and sees that they both have red crosses. So, naturally, she will put up her hand as she has been instructed to do. Similarly, B also sees two red crosses. So, she, too, raises her hand. C, likewise, sees two red crosses; and, of course, she, too, will raise her hand. At that point, A reasons as follows:

> *Let me assume that I have a blue cross on my forehead. If that is so, then one of the other two, say B, would know that she doesn't have a blue cross because otherwise C, seeing two blue crosses—mine and B's—would not have put up her hand. But this has not occurred. So, B and C cannot determine their color. This means that I too have a red cross.*

This type of puzzle is a model of how hypothesis thinking crystallizes in the mind and how it can be used to—literally—figure things out. Logic and imagination clearly form a blend in reasoning, or "reckoning," to use Ahmes' pertinent term.

Since antiquity, humanity has prided itself on being a *logical* species. Legend has it that the Greek philosopher Parmenides invented logic as he sat on a cliff meditating about the world. The English philosopher Thomas Hobbes (1656) claimed that logic was the only attribute that kept human beings from regressing to wild beasts—a view developed further by the French philosopher René Descartes, who refused to accept any belief, even the belief in his own existence, unless he could "prove" it to be logically true. Descartes (1637) also believed that logic was the only way to solve all human problems. In their insightful book, *Descartes' Dream*, Davis and Hersh (1986: 7) encapsulated Descartes' vision as "the dream of a universal method whereby all human problems, whether of science, law, or politics, could be worked out rationally, systematically, by logical computation." Leibniz (1646) forged the conceptual link between logic and life even further. Calling logical reasoning a *characteristica universalis* (a "universal language" of mind), Leibniz claimed that it could be used to great advantage in the betterment of the human condition for the simple reason that "errors" in thinking could be reduced to errors in logic and thus easily fixed.

As Rucker (1987: 218) aptly observes, the "great dream of rationalism has always been to find some ultimate theory that can explain *everything*." Perhaps what makes paradoxes so "mischievously appealing" is that they reveal ultimately why a "theory of everything" is impossible. They show that human systems, no matter how commonsensical they may appear to be, are ultimately imperfect. Gödel's theorem showed, in effect, that logic was made by imperfect logicians, and thus that the Cartesian dream of using logic to solve all human problems was illusory and ultimately empty. In anthropological terms, logic is a form of stable thinking, as argued throughout this chapter. After the imaginative work of the dreaming imagination, producing artifacts of mind based in *mythos*, the left hemisphere of the brain intervenes to give mnemonic form to the discoveries of the imagination. The blend of *mythos* and *lógos* is the neurological blueprint for the emergence of human cultures throughout the world.

7

Puzzles and Human Intelligence

Everything that is beautiful and noble is the product of reason and calculation.
Charles Baudelaire (1821–67)

Recall the Nine-Dot Puzzle (Chapter 1). It is a specific, or three-by-three version, of a general Dot-Joining Puzzle (as discussed). By solving sixteen-dot, twenty-five-dot, and various other puzzles, the question becomes: Is it possible to uncover some general abstract principle hidden within this type of puzzle? More specifically: Is there a correlation between number of dots and number of connecting lines?

By making the Dot-Joining Puzzle as complex as we desire (increasing the number of dots to 16, 25, 36, 49, and so on), a pattern seems to emerge through simple inspection. This pattern can be charted as shown in Table 7.1.

Table 7.1 Generalization of the Dot-Joining Puzzle

Dots	Lines Required
3 × 3	(3 + 1) = 4
4 × 4	(4 + 2) = 6
5 × 5	(5 + 3) = 8
6 × 6	(6 + 4) = 10
…	…
n × n	n + (n − 2) = 2n − 2 = 2(n − 1)

Research in recreational mathematics involves this kind of reasoning, corroborating indirectly the kind of flow model of puzzles put forth in this book (Kershaw and Ohlsson 2004). After the Aha insight has occurred in solving the original Nine-Dot puzzle, the reasoning part of the brain kicks in to impel us to explore logically if there are any recurrent patterns that may be packed into the puzzle. From this flow, characterized by a shift from imagination to reasoning, discoveries emanate in mathematics and other domains.

This final chapter aims to tie various thematic threads together, underscoring what puzzles reveal about the brain and why they are among the first creative and exploratory artifacts of human history, telling a fascinating neuro-anthropological story about the nature of human intelligence and why it is probably unique.

Homo ludens

Consider the common game of tic-tac-toe. As trivial as it might seem, it actually raises key questions related to abstract aspects of situational structure: What is the probability of winning if the X or the O is inserted in a particular location? What distribution makes sense at the start? Answering these questions entails the concurrent use of inferential and hypothesis thinking. In Figure 7.1, the X-player started the game and won.

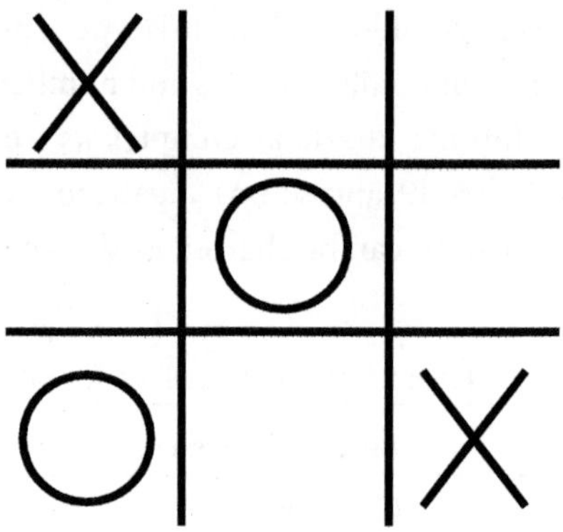

Figure 7.1 A game of tic-tac-toe.

Now, knowing the rules of the game and the fact that the X-player made the first placement, it is a fairly straight forward task to reconstruct the winning moves and explain them. Such reconstructions reveal how the imagination-reckoning neural blend occurs—we must envision the moves in reverse order and then map these against a placement system based on a series of related sequential moves. The game is an unconscious model of cognition that reveals, in its miniature way, how human intelligence works. Computer modeling of tic-tac-toe has revealed that there are 765 distinct game positions and over 26,000 possible games. As Moscovich (2015: 15) has perceptively remarked: "Despite its apparent simplicity, tic-tac-toe requires detailed analysis to determine even a few elementary combinatorial facts, like the number of possible positions." Interestingly, all games should produce a draw; so, error and miscalculations, along with one opponent outwitting the other, are the only ways for winning to occur. Mathematics does not fail; humans do, which is itself a significant lesson to learn from the world of puzzles, as has been emphasized in this book.

Such games are not only a means to gain recreation, but also a means for understanding ourselves. As Johan Huizinga (1938) has so aptly argued, *homo ludens* is an evolutionary twin of *homo sapiens*. Huizinga was actually criticized by claiming that ludic activity—mental and physical—is the main generator of culture. He modified this view subsequently, suggesting that ludic thinking is a necessary, albeit not sufficient, condition for culture. But his initial stance may not be a far-fetched one, given the ancientness and importance of puzzles to the foundations of culture, as argued throughout this book.

We play games and solve puzzles all the time, as if they were residues of some ancient archetypal form of cognition. Consider chess and all the puzzles that it has allowed over centuries. Take as a specific case in point the so-called *Eight Queens* puzzle, which was put forward in 1848 by Max Bezzel in the chess journal *Deutsche Schachzeitung*, whereby eight queens must be placed on an eight-by-eight chessboard in such a way that none of the queens is able to capture any other queen (with the normal rules of chess). A solution requires that no two queens share the same row, column, or diagonal. There are ninety-two distinct solutions to the puzzle, although if rotations and reflections of the board are taken into account, then it has twelve unique solutions. Now, this might seem like a contemporary engagement with mathematics, but it latently reveals that placing or moving objects in certain configurations in order to test what this entails is one of the most ancient of all mental activities: It is evident in Alcuin's River-Crossing Puzzles, in Euler's Königsberg Bridges Puzzle—and even in Sudoku. Moving symbols around to examine what it yields is also a guiding principle in anagrams, word squares, and the like. Moreover, it also exemplifies the presence of a flow structure to human intelligence. After the puzzle has been formulated, it leads to a generalization of its structure. The puzzle can, in fact, be solved with a recursive algorithm for placing "n queens" on an "n-by-n" chessboard. Simple algorithms have been constructed for various values of "n". These need not concern us here. The point is that it is a ludic artifact that, in its own miniature way, reverberates with latent implications for understanding human intelligence, which is a blend of imagination and reasoning, derived from observing the world in a ludic way.

The game of chess has, actually, been the basis for many puzzles that provide a template for analyzing ludic thinking. In any and all of its different cultural versions—for example, *shogi* in Japan and *xiang qi* in China (Bell and Cornelius

1988)—chess is played using specially designed pieces on a board of sixty-four alternating light and dark squares in eight rows of eight squares each. Given its outcomes, it has become, like tic-tac-toe, a model of human intelligence and its connection to human destiny—a fact that has not escaped the imagination of contemporary mathematicians and artists alike. For example, in Swedish director Ingmar Bergman's apocalyptic 1956 movie, *The Seventh Seal*, the narrative revolves around a game of chess between a knight and Death. The subtext of the movie is transparent—if, somehow, we were to understand the game of chess as a model of the game of life, then we could win the greatest game of all, *cheating Death*.

Raymond Smullyan's book, *The Chess Mysteries of Sherlock Holmes* (1979), is an ingenious compilation of chess puzzles, all of which tap into the archetypal nature of placement and its ability to model real-life situations. Appropriately enough, he presents his puzzles through a series of dialogues between master fictional sleuth Sherlock Holmes and his assistant Watson. Each puzzle requires the solver to grasp the structure of certain events in a game's past.

The latent power of *homo ludens* over cognitive activities today might be the reason why puzzles and games are popular and why "puzzle crazes" crystallize occasionally. Perhaps the most well-known craze is the Rubik's Cube, invented by the Hungarian professor of architecture, Ernö Rubik, in 1975. It was likely inspired by a popular game called the Thirty Colored Cubes Puzzle, invented in 1921 by P. A. MacMahon. Rubik taught at the Budapest School of Applied Arts, and claims to have invented his cube as an exercise in spatial reasoning for his students. By 1982, it had become an international craze, with 10 million sold in Hungary alone. As an aside, it should be mentioned that someone named Larry Nichols patented a similar puzzle cube in 1957 (Costello 1988: 148). In 1984, Nichols won a patent infringement lawsuit against the Ideal Toy Company, the American maker of the Rubik's Cube.

The puzzle is made up of smaller, colored cubes, so that each of the six faces of the large cube is a different color. The colors can be scrambled by twisting sections of the cube around any axis. The goal is to scramble the colors and then return the cube to its initial configuration, as is well known. Solutions typically consist of a sequence of moves that must be both envisioned and reasoned out in tandem. For example, one move might entail switching the locations of three corner pieces, while leaving the rest of the pieces in their places. Much research has been done on optimal solutions to the puzzle. The number of possible positions of the cube is given below (Frey and Singmaster 1982; Rubik 1987):

$$\frac{8!\ 12!\ 3^8 2^{12}}{2\cdot 3\cdot 2} = 43252003274489856000$$

It has been known since 1995 that a lower bound on the number of moves for the solution (in the worst case) was twenty, and it is now known that no configuration requires more than twenty moves. Various other mathematical implications have been derived from this game, including the fact that it manifests group structure. The cube thus not only taps into our need for ludic thinking, but also reveals how "flow thinking" is involved.

However, as we have seen previously, *homo ludens* is not only an experimenter with game structure, but also a trickster. Take Loyd's Fourteen/Fifteen Puzzle, which as the wily Loyd knew, is impossible to solve. Loyd put fifteen consecutively numbered sliding blocks in a square plastic tray large enough to hold sixteen such blocks. The blocks were arranged in numerical sequence, except for the last two, fourteen and fifteen, which were installed in reverse order. The object of the puzzle-game was to arrange the blocks into numerical sequence from one to fifteen, by sliding them, one at a time, into an empty square, without lifting any block out of the frame (Figure 7.2).

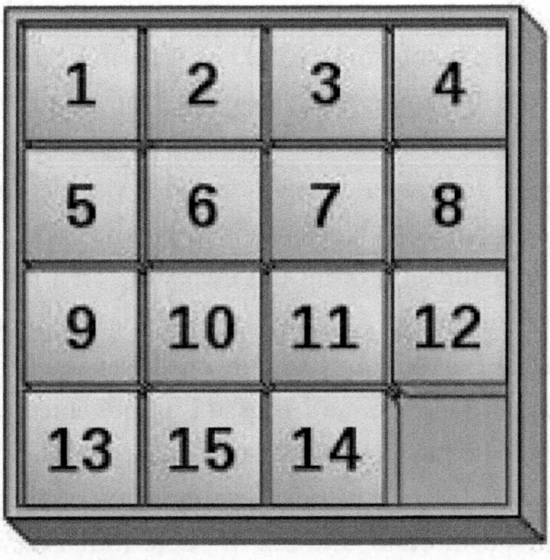

Figure 7.2 The Fourteen/Fifteen Puzzle (from Wikimedia Commons).

Although impossible to solve, it made the wily Loyd a considerable amount of money, nonetheless. People simply cannot ignore a ludic challenge, no matter what the costs are in time and energy. Incidentally, Loyd offered a prize of $1,000 for the first correct solution, knowing full well that the puzzle could never be solved. Notice that when the blocks are in numerical order, each one followed by a block that is exactly one digit higher (one followed by two, two by three, and so on), the blocks can be scrambled and the puzzle can be easily solved. This is called the "Fifteen Puzzle" (Figure 7.3).

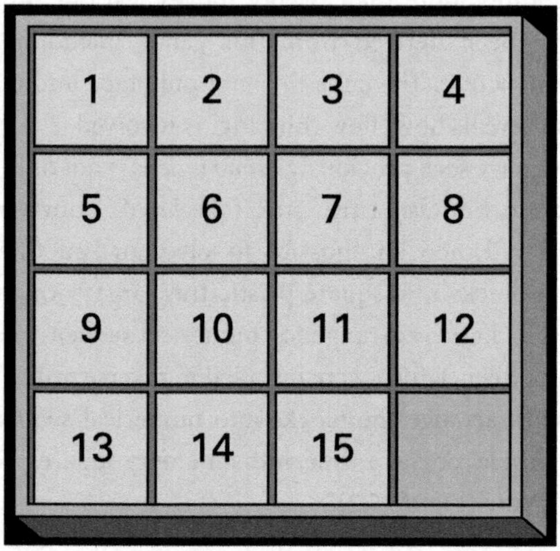

Figure 7.3 The Fifteen Puzzle (from Wikimedia Commons).

In any other arrangement, some blocks will be followed by blocks that are numerically lower (for example, two followed by one, four by three, etc.). Now, the remarkable thing is that the structure of this puzzle mirrors that of the Königsberg Bridges Puzzle, even though on the surface it does not seem to be related to it. Every instance of a block followed by one that is lower than itself is called an inversion. If the sum of all the inversions in a given arrangement is even, the solution to the puzzle is a possible one. If the sum is odd, it is impossible, just like Euler's puzzle. Loyd's game had only one inversion (fifteen is followed by fourteen). This is an odd number and, thus, it is impossible to arrange the blocks in numerical order. The connection to Euler's Bridges Puzzle, and even to Alcuin's River-Crossing Puzzles, is truly mind-boggling. Each of these puzzles requires moving from one position to another in determinable ways, whether the movement involves river crossings, interconnected bridges, or cell-to-cell shifts. In effect, the archetype is the same one; it simply manifests itself in different guises.

For the sake of historical accuracy, it should be mentioned there is doubt that Loyd was the actual inventor of this puzzle, and given his hucksterism, there is much credence assigned to this, as Bodycombe (2007: 13–14) elaborates:

> The truth is that he didn't invent it, and had nothing to do with popularizing the game in its heyday. The claim became the accepted truth and every obituary of Loyd included mention of "his" Fifteen Puzzle. So who did devise it? In 2006, a team led by puzzle historian Jerry Slocum managed to pin down the truth. An 1874 puzzle by a postmaster called Noyes Palmer Chapman was later copied by students at the American School for the Deaf, who began making the puzzle themselves. One of these puzzles was sold or given to a Boston-based woodworking shop owner, Matthias Rice, who repackaged it as the Gem Puzzle in December 1879. Various "me too" products followed when a reward for a correct solution was offered by a dentist called Charles Pevey in January 1880. The prize was $100 in cash plus a $25 set of teeth.

Puzzles and games that involve placement, movement, or combination are miniature models of how the human mind perceives objects and actions in the world, attempting to glean from these game-playing artifacts, insights into the structure of things. In 1957, Solomon W. Golomb invented the game of *polyominoes* as a brilliant contemporary version of this kind of ludic cognition (see Golomb 1965). A polyomino is a two-dimensional shape formed by joining square, flat pieces of paper or plastic. There are several basic types of polyominoes: the *domino* is a two-square figure, the *tromino* a three-square figure, the *tetromino*, a four-square figure, and the *pentomino* a five-square figure. Pentominoes can be assembled into a variety of forms.

In 1974, the British physicist Roger Penrose invented a different kind of assembly game, which has come to be called, appropriately, the Penrose Tile Game. It involves making *nonperiodic tilings* of the plane with figures. A *periodic tiling* is a design that recurs horizontally or vertically across the plane. The grid design of a sheet of graph paper is an example of such a tiling, since the same pattern of white square after white square recurs both vertically and horizontally. A *non-periodic tiling* does not recur either vertically or horizontally. Penrose's game produces an infinite number of different nonperiodic tilings of a surface, using two figures called a *dart* and a *kite*. Penrose used the golden ratio □ to design his darts and kites, connecting them to an archetype that has existed since antiquity at the very least. Penrose's tiles, like Golomb's polyominoes, are works of "ludic or puzzle art," inviting the eye to detect pattern and symmetry for their own sake. They produce, arguably, the same kind of aesthetic pleasure that comes from looking at a cubist painting.

Games of chance

Homo ludens seems also to be at the root of risky behaviors, since it impels the mind to engage in unknown things simply for the sake or enjoyment of it. But from the risks comes understanding. One of the most fascinating examples of this is in the realm of gambling. The foundations of modern-day probability theory, in fact, were laid by Girolamo Cardano, himself an avid gambler, in the sixteenth century. He was among the first to discuss and calculate the probability of throwing certain numbers with dice and of pulling certain cards, such as aces, from decks of cards. In his *Book of Games of Chance*, Cardano (reprinted 2015) illustrated how the dice and card games he played so frequently revealed a mathematical system of representation of the outcomes that had predictive value. Galileo Galilei also made relevant observations, based on his own game, *passadieci*, which the Grand Duke of Tuscany asked him to explain. Why is it that by throwing three dice the scores of ten and eleven occurred more often than nine and twelve? Galileo realized that the numbers ten and eleven can be obtained with more combinations of triplets, and so they are more likely to come up. Like Cardano, Galileo based his argument on the specification of the elementary events that will occur in an aleatory system, such as card and dice games. In the following century, Blaise Pascal and Pierre de Fermat developed these ideas into what came to be known as probability theory—hence, a new and unexpected branch of mathematics. The first formal text on the theory was Christian Huygens' treatise *De Ratiociniis in Ludo Alea* (1657). Huygens stated in the book that the seemingly trivial card and dice games harbored a "very interesting and deep theory." Huygens' treatise, also known as the first documented reference to the concept of mathematical expectation, was incorporated later by Jakob Bernoulli in his famous book *Ars Conjectandi*, in which he articulated the first limit theorem of probability theory—the law of large numbers (Borovkov 2013). It is to be emphasized that by examining seemingly trivial card and dice games, mathematicians such as Cardano, Galileo, Huygens, and Bernoulli founded probability theory.

To see in very concrete terms what this implies, consider a standard deck of fifty-two cards. Puzzles based on this deck abound and can be used to illustrate the connection of probability theory to other notions, such as permutations and combinations—hence emphasizing the connective nature of ludic thinking. Here is an example:

In how many different ways can four aces be drawn blindly from a standard deck?

First, we determine how many ways there are in drawing out any four cards. The answer is: 52 × 51 × 50 × 49 = 6,497,400. The reason is that any one of the fifty-two cards can be drawn first, of course. Each of the fifty-two possible first cards can be followed by any of the remaining fifty-one cards, drawn second. Since there are fifty-one possible second draws for each possible first draw, there are 52 × 51 possible ways to draw two cards from the deck. Now, for each draw of two cards, there are fifty cards left in the deck that could be drawn third. Altogether, there are 52 × 51 × 50 possible ways to draw three cards. Reasoning the same way, it is obvious that there are 52 × 51 × 50 × 49, or 6,497,400 possible ways to draw four cards from a standard deck.

To find the chances of getting the aces, it is next necessary to determine the number of four-ace draws there are among the 6,497,400 possible draws. We start by looking at each outcome, draw by draw. There are four aces that can be drawn first. For each one of these, there are three remaining aces that can be drawn second. Then, for each two-ace draw, there are two aces that can be drawn third. Finally, after three aces have been drawn, only one remains. So, the total number of four-ace arrangements is: 4 × 3 × 2 × 1 = 24. Thus, among the 6,497,400 ways to draw four cards there are twenty-four ways to draw them. The probability of doing so is, therefore, 24/6,497,400 =.0000036, which makes it a highly unlikely outcome.

This puzzle has several key features. First, it shows what probability theory is all about in a nutshell. Second, it shows that the methods of permutation and combination are connected, historically and practically, to probability theory. Third, it brings out concretely the relation between actual outcomes, visualization, and hypothetical thinking—the same form of thinking that underlies simple games like tic-tac-toe, as discussed.

Take, as another example, the following puzzle (Kasner and Newman 1940: 243–4):

> *Since there are four aces in a deck, the probability of drawing an ace from 52 cards is 4/52 = 1/13. But what is the probability of drawing either an ace or a king from the deck in one draw?*

This is an example of the probability of mutually exclusive or alternative events. The probability of drawing an ace is 1/13 and the probability of drawing a king is also 1/13. The probability of drawing either one or the other is the sum of these two: 1/13 + 1/13 = 2/13. In effect, there may be no better way to grasp this aspect of probability than through this, and similar, puzzles. Dice puzzles too can be used to illustrate various aspects of probability theory. The following is also from Kasner and Newman (1940: 244):

What is the probability of obtaining either a 6 or a 7 in throwing a pair of dice?

We start by listing the number of ways to throw either a six or a seven (Table 7.2).

Table 7.2 Outcomes of throwing two dice

Outcome: 6		**Outcome: 7**	
First Die	**Second Die**	**First Die**	**Second Die**
1	5	1	6
2	4	2	5
3	3	3	4
4	2	4	3
5	1	5	2
–	–	6	1

Now, there are thirty-six possible throws of two dice, because each of the six faces of the first die is matched with any of the six faces of the second one. Of these thirty-six possible throws, eleven produce either a six or a seven (as the table above shows). Therefore, the probability of throwing either a six or a seven is 11/36.

As these examples show, probability theory is essentially a means of assigning pattern to chance. It boggles the mind to contemplate that simple games can contain so many implications hidden within them. The whole idea of taming chance by "mathematizing" it into a theory reveals a desire to conquer uncertainty. But, as the French writer François de La Rochefoucauld argued in his *Maxims* (1665), this is foolish. Life is not predictable or predestined, let alone controllable. One's final destiny, La Rochefoucauld emphasized, will always constitute a disquieting psychic problem, no matter how many ingenious mathematical artifacts we create to control it. Probability theory, which is essentially a means of assigning mathematical *pattern* to *chance*, is, La Rochefoucauld suggested, an intellectual oxymoron. Yet, we continue to live through its cognitive reverberations.

Human intelligence

As discussed throughout this book, the key cognitive aspect of puzzles is that they require imaginative, or Aha, thinking, to various degrees. Even those that employ trickery, producing a Gotcha Effect, are still ensconced in the kind of thinking that triggers imaginative processes. Even a Sudoku puzzle requires that the solver imagine, or envision, number placements, combinations, and the like. When the diagram does not concede a definitive placement pattern, then the imagination suggests that we use hypothetical placements and from there make appropriate deductions. Thus, reasoning and imagination are blended to produce the relevant insight. The fact that the imagination is involved in the solution of puzzles is what has led many psychologists to equate Aha thinking with mental imagery and visualization. Actually, it is a matter of gradation. If puzzles are, indeed, solved on the basis of blended cognition, then it is perhaps more appropriate to think of this state of mind as locatable on a scale, ranging from zero or a low level of such thinking to one, a maximum level.

So, a particular puzzle can be seen to fall on this hypothetical scale. The closer it is to "1", the more it requires activating the imagination; the closer it is to "0", the more routine systematic thinking is involved. The Nine-Dot Puzzle can certainly be located near the 1-point end, whereas many Sudoku puzzles fall nearer to the 0-end. Other kinds of puzzles fall somewhere in between. Puzzles can be characterized as "miniature blueprints" of how blended thinking is wired into the brain and of how it allows us to come up spontaneously with solutions to intractable conundrums. Puzzles are, in a way, experiments in human intelligence and what it can do for us.

The ideas of Immanuel Kant (2011) are relevant in this regard. Kant described imaginative thinking in mathematics as "combining and comparing given concepts of magnitudes, which are clear and certain, with a view to establishing what can be inferred from them." Kant argued further that the whole process becomes reflective and explicit when we examine the "visible signs" that we use to highlight the structural details inherent in this type of knowledge. For example, a diagram of a triangle compared to that of a square will show where the differentiation occurs—one consists of three intersecting lines, while the other has four parallel and equal sides that form a boundary. This type of visual know-how is based on the brain's ability to synthesize scattered bits of information into holistic entities that can then be analyzed reflectively.

To see how Kant's description fits in with puzzle-solving, consider the following one by Diophantus, which he formulated in his *Arithmetica* (Diophantus 1910):

Divide a given number into two squares.

At first consideration, this seems to entail "brute force" logic. Now, while the trial-and-error method is sometimes part of deriving a solution, it is something that mathematicians wish to avoid in order to make the process efficient. Moreover, sometimes we cannot say for sure that trial-and-error procedures can lead to a true solution. Mathematicians always want to demonstrate that something is the way it is beyond any shadow of a doubt; extrapolating a solution from a given information, on the other hand, can only suggest that something is the way it is, not demonstrate it.

So, let's attack Diophantus' puzzle with a frame of mind that seeks to flesh out from it some hidden principle—the essence of blended thinking. The technique of setting an equation comes to mind to solve this in an easy and determinate fashion. The result is a quadratic equation, with the numbers being, 16, 16/5, and 12/5, so that $16 = (16/5)^2 + (12/5)^2$. The question becomes: What kind of thinking did Diophantus use? In this case, we actually have his solution in print (cited in Bashmakova 1997: 24–5):

> Let the first summand be x^2, and thus the second, $16–x^2$. The latter is to be a square. I form the square of the difference of an arbitrary multiple x diminished by the root of 16, that is, diminished by 4. I form, for example, the square of 2x–4. It is $4x^2 + 16 - 16x$. I put this expression equal to $16–x^2$. I add to both sides $x^2 + 16x$ and subtract 16. In this way I obtain $5x^2 = 16x$, hence $x = 16/5$. Thus one number is 256/25 and the other 144/25. The sum of these numbers is 16 and each summand is a square.

Diophantus' words reveal how his mind grappled with the puzzle. The Aha insight comes from seeing the problem as a general one involving quadratic equations. This kind of thinking involves what Kant called visible signs. The starting point is, in fact, an Aha moment that leads to a visualization of the appropriate signs. It is not surprising to find, by the way, that a similar phrase to "Aha" (in Egyptian, of course) is found in the *Ahmes Papyrus* (Spalinger 1990).

Now, consider the following puzzle devised by Lewis Carroll (1958b: 7):

> *A bag contains one counter, known to be either white or black. A white counter is put in, the bag shaken, and a counter drawn out, which proves to be white. What is now the chance of drawing a white counter?*

We let B and W–1 stand, respectively, for the black and white counters that might be in the bag at the start, and W–2 for the white counter added to the bag. The Aha insight is the fact that removing a white counter from the bag entails three equally likely combinations of two counters, one inside and one outside the bag:

Inside the Bag	*Outside the Bag*
1. W–1	W–2
2. W–2	W–1
3. B	W–2

In combination (1), the white counter drawn out is the one that was put into the bag (W–2), and the white counter inside it (W–1) is the counter originally in it. Combination (2) is the converse of (1): the white counter drawn out is the one that was originally in the bag (W–1), and the white counter inside it (W–2) is the counter that was put in. In combination (3), the white counter drawn out is the one that was put in the bag (W–2), since there was no white counter originally in it. The counter that remains in the bag is a black one (B). In two of the three cases, Carroll observed, a white counter remains in the bag. So, the chance of drawing a white counter on the second draw is two out of three.

This is unexpected, but easily understood because it taps into a common intelligence that is understood by anyone anywhere. Carroll's puzzle literally zeroes in on Aha thinking in progress. It outlines the main features of the puzzle situation and then allows the mind to extrapolate a solution from them. In other words, it shows us that the parts relate to each other in specific ways.

The solution process is complete after the insight produced in this way is formalized in some way. The hunch activates the imagination which, in turn, leads to some visual strategy or simply "visualization" (paraphrasing Kant), which leads, in turn, to the Aha insight. This is then given some form (equation, diagram, and so on) that suggests a simple solution. Finally, the experience of solving the puzzle is given a logical form through representation with signs and symbols. As one more example of how this kind of thinking unfolds, consider a famous probability puzzle, known as the Birthday Puzzle, which seems to produce a counter-intuitive result (Ball 1972):

How many people do there need to be in a room so that there is at least a 50% chance that two of them will share the same birthday?

The relevant hunch-insight is to ask the same question for different numbers of people, calculating the relevant probabilities, up till when the probability first drops below fifty percent. So, let's suppose that there are two people in a room. The total number of possible arrangements of birthdays in this case is:

365×365

If the two people do, indeed, have different birthdays, then the first one, say A, may have his or her on any day of the year (365 possibilities) and the second one, say B, may have his or hers on any day except the day of A's birthday. Thus, there are 364 possibilities for B's birthday. The number of possible pairs of distinct birthdays is given by the following formula:

365×364

The probability of one occurring is:

$$\frac{364 \times 365}{365 \times 365}$$

Now, we can generalize this approach to *n* people. In this case, the number of possible birthday arrangements is:

$$365^{n}$$

Assuming that every single person has a different birthday, then the same reasoning applies: A may have it on any of 365 days; B, on any of the remaining 364 days; C, on any of the then remaining 363 days; and so on, until the last, or n^{th} person, will have his or her birthday on any of the remaining $(365 - n)$ days in order to avoid the first $(n - 1)$ possibilities. The probability is:

$$\frac{364!}{(365 - n)! \times 365^{n-1}}$$

The first value for *n* for which this formula is below 0.5 is twenty-three. This means that twenty-three people will do the trick—a truly remarkable finding. More technically, the number of ways of assigning birthdays to people is the size of the set of functions from people to birthdays. The Birthday Puzzle reflects Kant's dictum that cognition involves "combining and comparing given concepts."

There is really no better way to explain the chain of discoveries that are triggered by a seeming simple or trivial puzzle concept other than to assume that they are the brain's own investigative tools. So, the River-Crossing Puzzle became a math meme that spread subsequently throughout the world of mathematics leading to combinatorics, critical path theory, and other branches. One can still see the echoes of Alcuin's puzzle reverberating in these complex systems. The same can be said about Euler's Königsberg Bridges Puzzle, Fibonacci's Rabbit Puzzle, and so on. As we saw with the Ahmes Papyrus and Alcuin's *Propositiones*, mathematicians have always understood that there is no better way to impart knowledge of mathematics than through puzzle formats. These stimulate interest and allow us to reconstruct or deconstruct the whole thinking process involved in solving them.

In a sentence, it would seem that for the brain to take in unfamiliar information it requires the experiential (probing) right-hemisphere functions to operate freely; these can be called R-Mode functions (Danesi 2003). However, this exploratory effort on the part of puzzle solvers would be virtually wasted if not followed up by the "analytical" intervention, so to speak, of the left hemisphere. This capacity can be called an L-Mode function. So, during the initial stages of solving puzzles, we assimilate input through creative R-Mode experiential activities. But after this stage, L-Mode thinking takes over to complement these stages. In effect, the R-Mode is a crucial point-of-departure for activating the appropriate imaginative areas of the brain. This is because the right hemisphere is an effective distributor of new information. It is an *efferent* hemisphere, designed anatomically to transmit impulses from its various neuronal networks to others throughout the brain. The left hemisphere, on the other hand, with its more sequentially organized neuronal-synaptic structure, is an *afferent* hemisphere, designed anatomically to receive impulses rather than transmit them.

Throughout this book, there has been an underlying subtext—namely, that when a solution comes seemingly spontaneously, it evokes great satisfaction and pleasure—the Aha Effect. Obviously, this is an area of research within psychology and even anthropology that needs much more serious attention. In his intriguing book *Aha! Insight!* Martin Gardner (1979) exemplifies how Aha effects are produced by actual puzzles and what they do to us emotionally and intellectually.

The blend of imagination and reasoning is likely the neural system that produces the "dialectic mind," as it has been called here. And this takes us back to the start of this foray into puzzles as dialectic constructs—that is, as products of the *fantasia* seeking answers to questions of existence. Puzzles are miniature expressions of this state of mind, reflecting its structure in their very form. Puzzle-solving, therefore, is not solely fortuitous or based on some special kind of intelligence as traditionally conceived. It comes about through a form of intuitive or imaginative thinking that is guided by the experience of recurring patterns in the world.

This kind of dialectic thinking is critical to the development of human knowledge. It is not a stretch to claim that it undergirds the foundations of culture—a form of communal living that seeks to solve the problems of existence by organizing them in specific ways. It was Socrates who turned dialectic reasoning concretely into the method of dialogue. There is no need here to go into the history and philosophy of the dialectic-dialogue nexus. The dialectic-ludic mind is innate in all humans—it consists in the need to ask questions and search for the answers to these very questions.

Mind and culture

Interest in the nature of the mind is as old as recorded history. The Greek historian Herodotus claimed, in his *Historia*, that Egyptians thought differently from Greeks because they wrote from right to left, rather than from left to right, as did the Greeks. Herodotus clearly understood two implicit things with this assertion: First, access to the mind is via its artifacts, such as writing; and second, the mind is not a rigid entity, it is shaped by environment and experiential events. A similar view was articulated by the fourteenth-century Algerian scholar Ibn Khaldun, who wrote a truly fascinating treatise, in which he noted that the subtle behavioral differences that existed between nomadic and city-dwelling Bedouins were due to their differences in vocabulary and in how they used them to represent and think about reality (Perron and Danesi 1999).

These perspectives prefigured Giambattista Vico's "poetic" approach to the study of mind (Bergin and Fisch 1984), whereby each human artifact can be considered a poetic form that contains the mind's thought imprints. Jumping forward to the present day, the same kind of perspective is mirrored in the notion of *autopoiesis*—a term introduced by Maturana and Varela in their famous 1973 book, *Autopoiesis and Cognition*, where they claimed that an organism participates in its own evolution, since it has the ability to produce, or at least shape, its various biochemical agents and structures, thus ensuring their efficient and economical operation. In the case of the human mind, autopoiesis seems to know no bounds. Like Vico's concept of *fantasia*, autopoiesis is an acknowledgment of the infinite and flexible capacity of human intelligence to produce and reproduce knowledge and insights in its own creative ways. In other words, we understand what we ourselves have made, as we attempt to develop models of the world. This is known generally as the *verum-factum* principle—*verum esse ipsum factum* ("truth itself is a fact" or "the truth is made"), which Vico applied throughout his *New Science*. As he puts it (Bergin and Fisch: 331): "The world of civil society has certainly been made by men, and its principles are therefore to be found within the modifications of our own human mind."

Puzzle-making throughout the ages is a product of the *verum-factum* principle. In his concluding remarks to his *New Science*, Vico turned specifically to the question of Divine Providence. He suggested that, although human beings may have created history and knowledge themselves, they were guided unwittingly in their efforts by a higher form of consciousness:

> It is true that men have themselves made this world of nations. . .but this world without doubt has issued from a kind often diverse, at times contrary, and always superior to the particular ends that men had proposed to themselves; which narrow ends, made means to serve wider ends, it has always employed to preserve the human race upon this earth.
>
> (Bergin and Fisch 1984: 1108)

As Fletcher (1991: 149) has observed, Vico's conceptualization prefigures much of modern cognitive science and anthropology because "it identifies *homo sapiens* with *homo faber*—cultural thinking with cultural making." To this can be added *homo ludens*. When groups congregate in a culture, they instantly start to think as one mind, which Vico called "common sense" (*sensus communis*). This is judgment without reflection, "shared by an entire class, an entire people, an entire nation, or the entire human race" (Bergin and Fisch 1984: 142).

Sensus communis comes from the *fantasia*, and its ability to connect concepts via ludic artifacts (riddles, games, placement puzzles, and so on). These reveal how thoughts originate in the mind—via experiential or affective connections that are felt to be related. So, the *fantasia* endows humans with the capacity to imagine the features of the world as one dynamic system. As Verene (1981: 101) puts it, the *fantasia* allows humans "to know from the inside" by extending "what is made to appear from sensation beyond the unit of its appearance and to have it enter into connection with all else that is made by the mind from sensation." Like the biologist Jakob von Uexküll (1909), Vico found a constant point of contact between the human body and consciousness. Von Uexküll argued that every species had different inward and outward lives. The key to understanding this duality could be found in the anatomical structure of the species itself and in the kind of brain it possessed. Animals with widely divergent anatomies and brains do not have access to the same kinds of experiences and perceptions. The *fantasia* is the source of autopoiesis, or self-construction, displacing humanity from the blind forces of biology; in effect, the human species has come to be regulated not only by force of natural selection *(il certo* or "certainty"), but also by the "force of history" (*il vero* "our truth").

Psychologists and anthropologists became interested in Vico starting in the 1980s because of a paradigm shift that was taking place in their disciplines at the time—an era when conceptual metaphor theory, as we saw, came into being. It is within this interdisciplinary frame of inquiry, based on blended thinking, that enigmatology fits in. As Northrop Frye (1990: xxii–xxiii) puts it, in true Vichian style, metaphor is the originating force of consciousness:

> The images of higher and lower, the categories of beauty and ugliness, the feelings of love and hatred, the associations of sense experience, can be expressed only by metaphor and yet cannot be either dismissed or reduced to projections of something else. Ordinary consciousness is so possessed by either–or contrasts of subject and object that it finds difficulty in taking in the notion of an order of words that is neither subjective nor objective, though it interpenetrates with

both. But its presence gives a very different appearance to many elements of human life, including religion, which depend on metaphor but do not become less "real" or "true" by doing so.

A similar purview of cognition can be found in so-called *phenomenology*, originally developed by the German philosopher Edmund Husserl (1891), to emphasize the importance of unconscious experience in the construction of categories of expression, from literature to science and, we might add, riddles, puzzles, and games. Husserl aimed to understand how awareness of sensations and emotions unfold in tandem with rational processes. Modern-day phenomenology characterizes the forms of consciousness as *phenomena*, and the processes involved in consciousness formation, such as perception and desiring, as *acts*. These are related to objects of consciousness and thus are also considered to be phenomena themselves. This intrinsic relationship between phenomena and acts is intentionality. Phenomenologists also claim that past experiences will limit people's ability to understand phenomena and thus to act accordingly. One way to counteract this limiting tendency of past experience is through what they call "fantasy variations;" that is, through acts of imagining how the same experience might be perceived under varying circumstances. The features of the experience that remain constant, despite the variations, are thus seen to constitute its essence. Husserl has had a number followers, even though it is not clear what the influence of phenomenology in the cognitive sciences has been, other than an emphasis on experience, which is now part and parcel of most psychological schools of thought. The followers include the French psychologist Maurice Merleau-Ponty (1942) and the German philosopher Martin Heidegger (1976). Both have argued that phenomenology should not be limited to an analysis of consciousness.

Children, like the first sentient humans, make truths through their imaginations (Bergin and Fisch 1984: 204–10, 403, 809, 933–4). At first, what will eventually become an abstract concept is formed as a god or a hero. Vico says of the ancient Egyptians that they could not form the concept of "civil wisdom." So, they formed it as thrice-Great Hermes (Bergin and Fisch 1984: 209). The ancient Greeks could not form the concept "valor." So, they formed what they meant by it through the character of Achilles (Bergin and Fisch 1984: 403). Children acquire concepts in identical ways—through god-like and heroic story characters who embody them. These are metaphors profiling human character, actions, behaviors, morals, and ethics. Metaphors, therefore, always state perceived truths: "the first founders of humanity applied themselves to sensory topics, by which they brought together those properties or qualities or relations of individuals and species which were, so to speak, concrete, and created from poetic genera" (Bergin and Fisch 1984: 495). For this reason, "metaphor makes up the great body of language among all nations" (Bergin and Fisch 1984: 444), allowing us to give "names to things from the most particular and the most sensible ideas" (Bergin and Fisch 1984: 442). If we change the word metaphor to riddle or puzzle, the foregoing discussion summarizes the overarching theme of this book. Puzzles are metaphors in a broad philosophical sense—they blend different domains of thought into unitary wholes. As Howard Gardner (1982: 161) observes, metaphor allows the mind "to perceive a resemblance between elements in two separate domains or areas of experience and to link them together in linguistic form."

Today, a new stream in psychology, known as evolutionary psychology (EP), has emerged. At first blush, it appears to have reverberations with Vichian thought, since it directly connects cognition to bodily based mechanisms. It also seems to fit into the model of dialectic-ludic intelligence discussed here. But it is really a modern form of determinism, standing in opposition to the autopoietic movement discussed above. EP focuses on genetic processes as generative of cultural ones. It is a persuasive form of discourse that fits in with the current tendency in some domains of cognitive science to explain consciousness in strictly genetic, rather than historical-cultural-imaginative, terms (Buss 2004). EP basically sees the mind as having the same modular structure of the body, with different adaptations serving different functions. EP is based on several core premises:

1. The brain is an information-processing device that translates external inputs into behavioral outputs.
2. The brain's adaptive mechanisms are the result of natural and sexual selection.
3. The brain has evolved specialized mechanisms to solve recurring problems of survival, which have remained part of the triune brain.
4. Most neural processes are unconscious and these enter automatically into the resolution of the problems.

EP is actually an offshoot of sociobiology. The key figure behind it is the American biologist Edward O. Wilson (1975, 1979, 1984), known for his work tracing the effects of natural selection on biological communities, especially on populations of insects, and for extending the idea of natural selection to human cultures. Since the mid-1950s, Wilson has maintained that social behaviors in humans are genetically based and that evolutionary processes favor those behaviors that enhance reproductive success and survival. Thus, characteristics such as heroism, altruism, aggressiveness, and male dominance should be understood as evolutionary outcomes, not in terms of social or psychic processes. Moreover, Wilson sees the creative capacities undergirding language, art, scientific thinking, myth, etc. as originating in genetic responses that help the human organism survive and continue the species. As he has stated rather bluntly, "no matter how far culture may take us, the genes have culture on a leash" (Wilson and Harris 1981: 464).

Conscious social behaviors are, of course, partially based in biology; but they are not totally so. Genetic factors alone do not completely define human beings. They tell us nothing about why humans create their meaningful experiences and pose the questions they do about life. In his brilliant book, *Mental Models* (1983), Johnson-Laird provided a good classificatory matrix for talking about consciousness:

1. "Cartesian machines" which lack awareness of themselves.
2. "Craikian machines" (after Craik 1943) that have the capacity of constructing models of reality, but lack self-awareness.
3. Self-reflective machines that construct models of reality and are aware of their ability to construct such models.

Computer programs designed to simulate human intelligence are Cartesian machines, whereas animals and human infants are Craikian machines. But only human infants have the capacity to develop self-reflective consciousness through growth, which is autopoietic, capable of self-generation and self-maintenance. Self-aware machines are able to act and think intentionally rather than merely as if they were acting so (of which Craikian machines are capable). This is because they can create truth (*verum-factum*) and decide what kind of reality should come into being.

As the emergence of puzzles in human life suggest, we are, indeed, makers of truth, playing on it in various ways. *Homo faber* and *homo ludens* are partners with *homo sapiens*. Puzzles provide models of how these three evolutionary figures participate with each other in producing human intelligence, knowledge, and culture.

Concluding remarks

A tale that is told about Thomas Edison indicating that he would sit down in a comfortable chair with a metal ball in his hand when trying to figure out a tough problem. As he drifted off into sleep, the ball would fall, waking him, and he would awake with the solution to the problem, as it came to him in his dream state. This tale may or may not be true, but it encapsulates how many intractable puzzles are solved and even how they may have come about in the first place. There are, in fact, many stories of mathematicians solving difficult problems while dreaming. Henri Poincaré recounted a period of hard and seemingly fruitless effort to solve a problem. To take a break, he went on a geological expedition. As he was getting on a bus, he made the most important breakthrough of his life, with the solution coming to him out of nowhere.

The ancients certainly were aware of the power of dreams. The first acts of human intelligence were likely the product of unconscious dreams that were given linguistic form in mythic narratives and riddles. Because they seemed to descend from nowhere, the ancients believed that they were keys to unraveling "dark and obscure secrets," as Ahmes so poetically put it. The Pythagorean Brotherhood espoused the view that mathematics was the key to unraveling the "secrets of the universe" for the same reason. Pythagoras prescribed obedience, silence, fasting, simplicity in dress, minimal possessions, and self-examination, so that the greater goal of the quest for truth would not become derailed by frivolities. The Brotherhood's systematic investigations of numbers and their geometric properties gradually led to an autonomous discipline of mathematics, which they saw as a guide to decoding different phenomena inherent in the world. By connecting numbers and musical forms, for instance, the Pythagoreans sought to understand how everything in the world was intertwined—a view that has come to be known as the "Harmony of the Spheres."

Stuart Isacoff (2003) has argued that the invention of Western musical traditions came about from the Pythagorean legacy that the "natural structure" of music is mathematical and mirrors the harmony of the emotions in humans. Thus, the "secret" as to why we react to, say, the symphonies of Beethoven so fervently, is due to its mathematical symmetry. Ian Stewart (2008: 9) has eloquently summarized this view as follows:

> The main empirical support for the Pythagorean concept of a numerical universe comes from music, where they had noticed some remarkable connections between harmonious sounds and simple numerical facts. Using simple experiments they discovered that if a plucked string produces a note with a particular pitch, then a string half as long produces an extremely harmonious note, now called the octave. A string two-thirds as long produces the next most harmonious note, and one three-quarters as long also produces a harmonious note. These two numerical aspects of music are traced to the physics of vibrating strings, which move in patterns of waves. The number of waves that can fit into

a given length of string is a whole number, and these whole numbers determine the simple numerical ratios. If the numbers do not form a simple ratio then the corresponding notes interfere with each other, forming discordant "beats" which are unpleasant to the ear. The full story is more complex, involving what the brain is accustomed to, but there is a definite physical rationale behind the Pythagorean discovery.

Indirect support for Pythagoras' view came forward in 1865, when English chemist John Newlands arranged the elements according to atomic weight and discovered that those with similar properties occurred at every eighth element like the octaves of music. He called this finding, in fact, the Law of Octaves. In turn, it led to the development of the Periodic Law of chemical elements. But the Harmony of the Spheres view also revealed some unexpected and "unwanted" findings. When the sides of a right triangle are of unit length, then the hypotenuse turns out to be a non-integer value. This truly upset the harmonic view of the universe. Also, when ratios between certain string vibrations are set, other ratios are thrown off, thus producing dissonances. The Pythagoreans knew about these defects in their philosophy, but kept them secret. To banish the dissonances, the keyboard of the piano (clavier) was tempered by breaking the octave into equal parts, so that all harmonies sounded in tune—an event attributed to Johann Sebastian Bach's *Well-Tempered Clavier* (1742). It was a human invention that rectified the Pythagorean defect, thus establishing a musical tradition that continues to this day. This episode illustrates that discovery and invention may be two sides of the same coin, so to speak. The difference is that discovery is serendipitous, whereas invention is intentional. It also shows the connectivity among seemingly disparate phenomena or artifacts. It is the same connectivity that can be seen among all puzzle genres, which are based on imaginative hunches that become reflective thoughts in a cognitive flow that requires no particular training.

The Pythagorean approach to mathematics was, at its core, a search for intrinsic pattern in the universe. Puzzles, too, can be characterized as pattern-detecting devices. Enigmatology is essentially a study of the ludic brain via its artifacts. Lewis Carroll, perhaps the greatest puzzlist of all time, claimed that he meant only nonsense with his puzzles. But the type of nonsense to which Carroll alludes is the same one that motivated Ahmes, Alcuin, Archimedes, Fibonacci, Cardano, and Euler, to mention but a few. Puzzles are not just curious figments of the mind, but bits of evidence of an elusive theory of the world that is lurking around somewhere, but that seems to evade formulation. Originating at the dawn of civilization, puzzles are among the oldest products of human ingenuity. The great American writer Henry Miller once proclaimed that many of the seemingly "trivial things" that we have produced throughout history may exist to counteract a deeply rooted feeling within us that life may have no meaning. That is certainly true of puzzles. In their own miniature way, and as trivial as they may seem, puzzles fill an existential void that we would otherwise feel constantly within us by providing answers to questions. They are as intrinsic to human nature as are humor, language, art, music, and all of the other creative faculties that distinguish humanity from other species.

References

Aarne, Antti (1918). *Vergleichende Rätselforschungen*, 3 vols. Helsinki: Suomalainen Tiedeakatemia.

Abbott, Edwin A. (2002 [1884]). *The Annotated Flatland: A Romance of Many Dimensions*. Introduction and Notes by Ian Stewart. New York: Basic Books.

Aczel, Amir D. (2000). *The Mystery of the Aleph: Mathematics, the Kabbalah and the Search for Infinity*. New York: Four Walls Eight Windows.

Adam, John A. (2004). *Mathematics in Nature: Modeling Patterns in the Natural World*. Princeton, NJ: Princeton University Press.

Ahrens, W. (1921). *Mathematische Unterhalten und Spiele*. Leipzig: B. G. Teubner.

Al-Khalili, Jim (2012). *Paradox: The Nine Greatest Enigmas in Physics*. New York: Broadway.

Alberti, Leon Battista (1997). *A Treatise on Ciphers*, A. Zaccagnini (trans). Torino: Galimberti.

Aliseda, Atocha (2010). *Abductive Reasoning*. New York: Springer.

Amthor, A. and Krumbiegel, B. (1880). Das Problema Bovinum des Archimedes. *Zeitschrift für Mathematik und Physik* 25: 121–71.

Andrews, William S. (1960). *Magic Squares and Cubes*. New York: Dover.

Archimedes (1897). *The Sand Reckoner*, Thomas L. Heath (trans.). Cambridge: Cambridge University Press.

Aristotle (1952). *Rhetorica*. In: *The Works of Aristotle*, Vol. 11, W. D. Ross (ed.). Oxford: Clarendon Press.

Arnheim, Rudolf (1969). *Visual Thinking*. Berkeley, CA: University of California Press.

Asch, Solomon (1950). On the Use of Metaphor in the Description of Persons. In: *On Expressive Language*, pp. 86–94, H. Werner (ed.). Worcester, MA: Clark University Press.

Ascher, Marcia (1990). A River-Crossing Problem in Cross-Cultural Perspective. *Mathematics Magazine* 63: 26–9.

Atkinson, Donald (1951). The Origin and Date of the "Sator" Word Square. *Journal of Ecclesiastical History* 2: 1–18.

Bachet, Claude-Gaspar (1984). *Problèmes plaisans et délectables qui se font par les nombres*. Lyon: Gauthier-Villars.

Ball, W. W. Rouse (1972). *Mathematical Recreations and Essays*, 12th edition, revised by H. S. M. Coxeter. Toronto: University of Toronto Press.

Barrow, John (2014). *100 Essential Things You Didn't Know about Maths & the Arts*. London: Bodley Head.

Barthes, Roland (1957). *Mythologies*. Paris: Seuil.

Barwise, Jon and Etchemendy, John (1986). *The Liar*. Oxford: Oxford University Press.

Bashmakova, Izabella G. (1997). *Diophantus and Diophantine Equations*. Washington, DC: Mathematical Association of America.

Bell, Robbie C. and Cornelius, Michael (1988). *Board Games Round the World: A Resource Book for Mathematical Investigations*. Cambridge: Cambridge University Press.

Benjamin, Arthur, Chartrand, Gary, and Zhang, Ping (2015). *The Fascinating World of Graph Theory*. Princeton, NJ: Princeton University Press.

Benson, William H. and Jacoby, Oswald (1981). *Magic Cubes: New Recreations*. New York: Dover.

Bergin, Thomas G. and Fisch, Max (eds. and trans.) (1984). *The New Science of Giambattista Vico*. Ithaca, NY: Cornell University Press.

Berry, John W. (1968). Ecology, Perceptual Development and the Müller-Lyer Illusion. *British Journal of Psychology* 59: 205–10.

Black, Max (1962). *Models and Metaphors*. Ithaca, NY: Cornell University Press.

Blatner, David (1997). *The Joy of Pi*. Harmondsworth: Penguin.

Bodycombe, David J. (2007). *The Riddles of the Sphinx*. London: Penguin.

Bombaugh, Charles C. (1962). *Oddities and Curiosities of Words and Literature*. New York: Dover.

Boole, George (1854). *An Investigation of the Laws of Thought*. New York: Dover.

Borgmann, Dimitri A. (1965). *Language on Vacation*. New York: Scribner's.

Borovkov, Alexander A. (2013). *Probability Theory*. New York: Springer.

Bronowski, Jacob (1973). *The Ascent of Man*. Boston, MA: Little, Brown and Co.

Brooke, Maxey (1969). *150 Puzzles in Crypt-Arithmetic*. New York: Dover.

Brown, Roger W. (1958). *Words and Things*. New York: The Free Press.

Bruno, Giuseppe, Genovese, Andrea, and Improta, Gennaro (2013). Routing Problems: A Historical Perspective. In: *The Best Writing in Mathematics 2012, pp. 197–208*, M. Pitici (ed.). Princeton, NJ: Princeton University Press.

Bühler, Karl (1908 [1951]). On Thought Connection. In: *Organization and Pathology of Thought*, pp. 81–92, D. Rapaport (ed.). New York: Columbia University Press.

Bunt, Lucas N., Jones, Phillip S., and Bedient, Jack D. (1976). *The Historical Roots of Elementary Mathematics*. New York: Dover.

Burkholder, Peter J. (1993). Alcuin of York's *Propositiones ad acuendos juvenes*: Introduction, Commentary, and Translation. *History of Science & Technology Bulletin*, Vol. 1, No. 2.

Burrows, Hannah (2014). Wit and Wisdom: The Worldview of the Old Norse-Icelandic Riddles and Their Relationship to Eddic Poetry. In: *Eddic, Skaldic, and Beyond: Poetic Variety in Medieval Iceland and Norway*, pp. 114–35, Martin Chase (ed.). New York: Fordham University Press.

Buss, David M. (2004). *Evolutionary Psychology: The New Science of the Mind*. Boston, MA: Pearson.

Campbell, Joseph (1972). *Myths to Live By*. Harmondsworth: Penguin.

Cardano, Girolamo (2015). *The Book of Games of Chance*. New York: Dover.

Carroll, Lewis (1868). The Alphabet Cipher. www.christies.com/lotfinder/books-manuscripts/dodgson-charles-lutwidge-the-alphabet-cipher-london-5280733-details.aspx.

Carroll, Lewis (1880). *Pillow Problems and a Tangled Tale*. New York: Dover.

Carroll, Lewis (1886). *A Tangled Tale*. London: Macmillan.

Carroll, Lewis (1958a). *The Game of Logic*. New York: Dover.

Carroll, Lewis (1958b). *Mathematical Recreations of Lewis Carroll*. New York: Dover.

Céard, Jean and Margolin, J. C. (1986). *Rébus de la Renaissance: des images qui parlent*. Paris: Maisonneuve et Larose.

Chambers, R. W., Förster, Max, and Flower, Robin (1933). *The Exeter Book of Old English Poetry*. London: P. Lund.

Chase, Arnold B. (1979). *The Rhind Mathematical Papyrus: Free Translation and Commentary with Selected Photographs, Transcriptions, Transliterations and Literal Translations*. Reston, VA: National Council of Teachers of Mathematics.

Chevalier, Jean and Gheerbrant, Alain (1994). *The Penguin Dictionary of Symbols*. Harmondsworth: Penguin.

Clark, Ronald W. (1977). *The Man Who Broke Purple*. London: Weidenfeld & Nicolson.

Clawson, Calvin C. (1996). *Mathematical Mysteries: The Beauty and Magic of Numbers*. Cambridge, MA: Perseus Books.

Conrad, Axel, Hindrichs, Tanja, Morsy, Hussein, and Wegener, Ingo (1994). Solution of the Knight's Hamiltonian Path Problem on Chessboards. *Discrete Applied Mathematics* 50: 125–34.

Cook, William J. (2014). *In Pursuit of the Traveling Salesman Problem*. Princeton, NJ: Princeton University Press.

Costello, Matthew J. (1988). *The Greatest Puzzles of All Time*. New York: Dover.

Craik, Kenneth (1943). *The Nature of Explanation*. Cambridge: Cambridge University Press.

Cutler, Bill (2003). Solution to Archimedes' Loculus, www.billcutlerpuzzles.com.

D'Israeli, Isaac (1835). *Curiosities of Literature*. London: Edward Moxon.

Dalgety, James and Hordern, Edward (1999). Classification of Mechanical Puzzles and Physical Objects Related to Puzzles. In: *The Mathematician and Pied Puzzler: A Collection in Tribute to Martin Gardner*, pp. 175–86, Elwyn Berlekamp and Tom Rodgers (eds.). Natick, MA: A. K. Peters.

Danesi, Marcel (2002). *The Puzzle Instinct: The Meaning of Puzzles in Human Life*. Bloomington, IN: Indiana University Press.

Danesi, Marcel (2003). *Second Language Teaching: A View from the Right Side of the Brain*. Dordrecht: Kluwer.

Danesi, Marcel (2016). *Language and Mathematics: An Interdisciplinary Guide*. Berlin: Mouton de Gruyter.

Danesi, Marcel (2017). The Bidirectionality of Metaphor. *Poetics Today* 38: 15–33.

Danesi, Marcel (2018). What is a Puzzle? Five Classic Examples. *Psychology Today*, www.psychologytoday.com, March 17, 2018.

Darling, David (2004). *The Universal Book of Mathematics: From Abracadabra to Zeno's Paradoxes*. New York: John Wiley and Sons.

Davis, Paul J. and Hersh, Reuben (1986). *Descartes' Dream: The World According to Mathematics*. Boston, MA: Houghton Mifflin.

Dawkins, Richard (1976). *The Selfish Gene*. Oxford: Oxford University Press.

de Bono, Edward (1970). *Lateral Thinking: Creativity Step-by-Step*. New York: Harper & Row.

De Morgan, Augustus (1954). *A Budget of Paradoxes*. New York: Dover.

Degrazia, Joseph (1981). *Math Tricks, Brain Twisters & Puzzles*. New York: Bell Publishing Company.

Dehaene, Stanislas (1997). *The Number Sense: How the Mind Creates Mathematics*. Oxford: Oxford University Press.

Descartes, René (1637). *Essaies philosophiques*. Leyden: L'imprimerie de Ian Maire.

Devlin, Keith (1999). *Mathematics: The New Golden Age*. Harmondsworth: Penguin.

Devlin, Keith (2011). *The Man of Numbers: Fibonacci's Arithmetic Revolution*. New York: Walker and Company.

Diophantus (1910). *Diophantus of Alexandria*, Thomas L. Heath (trans.). Cambridge: Cambridge University Press.

Doob, Penelope Reed (1990). *The Idea of the Labyrinth*. Ithaca, NY: Cornell University Press.

Doody, Margaret (2015). *Jane Austen's Names: Riddles, Persons, Places*. Chicago, IL: University of Chicago Press.

Dooley, John F. (2016). Cryptology in Fiction, http://faculty.knox.edu/jdooley/Crypto/CryptoFictionByAuthor.htm.

Dudeney, Henry E. (1917). *Amusements in Mathematics*. New York: Dover.

Dudeney, Henry E. (1958). *The Canterbury Puzzles and Other Curious Problems*. New York: Dover.

Dundes, Alan (1963). Towards a Structural Definition of the Riddle. *Journal of American Folklore* 76: 111–18.

Durkheim, Émile (1912). *The Elementary Forms of Religious Life*. Paris: Presses Universitaires de France.

Eckler, Ross (1997). *Making the Alphabet Dance: Recreational Wordplay*. New York: St. Martin's Griffin.

Eco, Umberto (1984). *Semiotics and the Philosophy of Language*. Bloomington, IN: Indiana University Press.

Elwes, Richard (2014). *Mathematics 1001*. Buffalo, NY: Firefly.

Fauconnier, George and Turner, Mark (2002). *The Way We Think: Conceptual Blending and the Mind's Hidden Complexities*. New York: Basic.

Fibonacci, Leonardo (2002). *Liber Abaci*, L. E. Sigler (trans.). New York: Springer.

Fisher, Adrian and Gerster, Georg (1990). *The Art of the Maze*. London: George Weidenfeld & Nicolson.

Fletcher, Angua (1991). *Colors of the Mind: Connections on Thinking in Literature.* Cambridge, MA: Harvard University Press.

Fortnow, Lance (2013). *The Golden Ticket: P, NP, and the Search for the Impossible.* Princeton, NJ: Princeton University Press.

Franksen, Ole I. (1985). *Mr. Babbage's Secret: The Tale of a Cipher—and APL.* Englewood Cliffs, NJ: Prentice Hall.

Frege, Gottlob (1879). *Begiffsschrift eine der Aritmetischen nachgebildete Formelsprache des reinen Denkens.* Halle: Nebert.

Freiberger, Marianne (2006). Flatland: A Review. *Plus*, https://plus.maths.org/content/flatland, accessed May 4, 2017.

Freud, Sigmund (1901). *The Interpretation of Dreams.* New York: Avon.

Frey, Alexander H. and Singmaster, David (1982). *Handbook of Cubic Math.* Hillside, NJ: Enslow.

Frye, Northrop (1981). *The Great Code: The Bible and Literature.* Toronto: Academic Press.

Gaines, Helen F. (1956). *Cryptanalysis.* New York: Dover.

Gardner, Martin (1961). *The 2nd Scientific American Book of Mathematical Puzzles.* New York: Simon and Schuster.

Gardner, Martin (1970). The Fantastic Combinations of John Conway's New Solitaire Game "Life." *Scientific American* 223: 120–3.

Gardner, Martin (1972). *Codes, Ciphers and Secret Writing.* New York: Dover.

Gardner, Martin (1979). *Aha! Insight!* New York: Scientific American.

Gardner, Martin (1982). *Gotcha! Paradoxes to Puzzle and Delight*, San Francisco, CA: Freeman.

Gardner, Martin (1994). *My Best Mathematical and Logic Puzzles.* New York: Dover.

Gardner, Martin (1997). *The Last Recreations: Hydras, Eggs, and Other Mathematical Mystifications.* New York: Copernicus.

Gillings, Richard J. (1961). Think-of-a-Number: Problems 28 and 29 of the Rhind Mathematical Papyrus. *The Mathematics Teacher* 54: 97–102.

Gillings, Richard J. (1962). Problems 1 to 6 of the Rhind Mathematical Papyrus. *The Mathematics Teacher* 55: 61–5.

Gillings, Richard J. (1972). *Mathematics in the Time of the Pharaohs.* Cambridge, MA: MIT Press.

Gleick, James (1987). *Chaos: The Making of a New Science.* New York: Viking.

Gödel, Kurt (1931). Über formal unentscheidbare Sätze der Principia Mathematica und verwandter Systeme, Teil I. *Monatshefte für Mathematik und Physik* 38: 173–89.

Golomb, Solomon W. (1965). *Polyominoes.* New York: Scribners.

Gullberg, Jan (1997). *Mathematics from the Birth of Numbers.* New York: W. W. Norton.

Hadley, John and Singmaster, David (1992). Problems to Sharpen the Young. *Mathematics Gazette* 76: 102–26.

Haken, Wolfgang (1977). Every Planar Map is Four-Colorable. *Illinois Journal of Mathematics* 21: 429–567.

Haken, Wolfgang and Appel, Kenneth (1977). The Solution of the Four-Color-Map Problem. *Scientific American* 237: 108–21.

Haken, Wolfgang and Appel, Kenneth (2002). The Four-Color Problem. In: *Philosophy of Mathematics*, pp. 193–208, D. Jacquette (ed.). Oxford: Blackwell.

Hannas, Linda (1972). *The English Jigsaw Puzzle, 1760–1890*. London: Wayland.

Hannas, Linda (1981). *The Jigsaw Book: Celebrating Two Centuries of Jigsaw-Puzzling Round the World*. New York: Dial.

Hardy, G. H. (1967). *A Mathematician's Apology*. Cambridge: Cambridge University Press.

Heesterman, J. C. (1997). On Riddles. In: *Significations: Essays in Honour of Henry Schogt*, pp. 65–9, P. Bhatt (ed.). Toronto: Canadian Scholars' Press.

Heidegger, Martin (1976). *Phaenomenologie und Theologie*. Frankfurt: Klostermann.

Herodotus (440 BCE). *The Histories*. Library of Alexandria 2014.

Hersh, Reuben (1998). *What is Mathematics, Really?* Oxford: Oxford University Press.

Hobbes, Thomas (1656). *Elements of Philosophy*. London: Molesworth.

Holroyd, Stuart and Powell, Neil (1991). *Mysteries of Magic*. London: Bloomsbury.

Honeck, Richard P. and Hoffman, Robert R. (eds.) (1980). *Cognition and Figurative Language*. Hillsdale, NJ: Lawrence Erlbaum Associates.

Hooke, S. H. (1935). *The Labyrinth*. London: Society for Promoting Christian Knowledge.

Hooper, William (1782). *Rational Recreations*. London: L. Davis.

Hovanec, Helene (1978). *The Puzzlers' Paradise: From the Garden of Eden to the Computer Age*. New York: Paddington Press.

Hsiung, Ming (2009). Jump Liars and Jourdain's Card via the Relativized T-Scheme. *Studia Logica* 91: 239–71.

Huizinga, Johan (1938). *Homo Ludens: A Study of the Play-Element in Human Culture*. New York: Beacon Press.

Hunter, James A. H. (1965). *Fun with Figures*. New York: Dover.

Huntley, H. E. (1970). *The Divine Proportion*. New York: Dover.

Husserl, Edmund (1970 [1891]). *Philosophie der Arithmetik*. The Hague: Nijhoff.

Huxley, Aldous (1928). Ravens and Writing Desks. *Vanity Fair*, September 1928.

Huygens, Christiaan (1657 [2010]). *Of the Laws of Chance*, trans. Ecco Publications.

Isacoff, Stuart (2003). *Temperament: How Music Became a Battleground for the Great Minds of Western Civilization*. New York: Knopf.

Jaynes, Julian (1976). *The Origins of Consciousness and the Breakdown of the Bicameral Mind*. Toronto: University of Toronto Press.

Johnson, Samuel (1755 [2007]). *A Dictionary of the English Language*. Harmondsworth: Penguin.

Johnson, Steven (2005). *Everything Bad is Good for You: How Popular Culture is Making Us Smarter*. New York: Riverside Books.

Johnson-Laird, Philip N. (1983). *Mental Models*. Cambridge, MA: Harvard University Press.

Jung, Carl G. (1983). *The Essential Jung*. Princeton, NJ: Princeton University Press.

Kaivola-Bregenhøj, Annikki (2001). Riddles: Perspectives on the Use, Function, and Change in a Folklore Genre. *Studia Fennica, Folkloristica* 10: 11–12.

Kant, Immanuel (2011). *Critique of Pure Reason*, J. M. D. Meiklejohn (trans.). CreateSpace Platform.

Kappraff, Jay (1991). *Connections: The Geometric Bridge between Art and Science*. New York: McGraw-Hill.

Kasner, Edward and Newman, James R. (1940). *Mathematics and the Imagination*. New York: Simon and Schuster.

Kershaw, Trina and Ohlsson, Stellan (2004). Multiple Causes of Difficulty in Insight: The Case of the Nine-Dot Problem. *Journal of Experimental Psychology: Learning, Memory, and Cognition* 30: 3–13.

Kim, Scott (2016). *What is a Puzzle?* scottkim.com.previewc40.carrierzone.com/thinkinggames/whatisapuzzle.

Kiryushchenko, Vitaly. (2012). The Visual and the Virtual in Theory, Life and Scientific Practice: The Case of Peirce's Quincuncial Map Projection. In: *Semiotic and Cognitive Science Essays on the Nature of Mathematics*, pp. 61–70, Mariana Bockarova, Marcel Danesi, and Rafael Núñez (eds.). Munich: Lincom Europa.

Kosslyn, Stephen M. (1983). *Ghosts in the Mind's Machine: Creating and Using Images in the Brain*. New York: W. W. Norton.

Kosslyn, Stephen M. (1994). *Image and Brain*. Cambridge, MA: MIT Press.

Kraitchik, Maurice (1942). *Mathematical Recreations*. New York: W. W. Norton.

Kuhn, Thomas S. (1970). *The Structure of Scientific Revolutions*. Chicago, IL: University of Chicago Press.

La Rochefoucauld, François de (1657 [2006]). *Maxims*. New York: Dover.

Lakoff, George and Johnson, Mark (1980). *Metaphors We Live By*. Chicago, IL: Chicago University Press.

Lakoff, George and Núñez, Rafael (2000). *Where Mathematics Comes From: How the Embodied Mind Brings Mathematics into Being*. New York: Basic Books.

Lapidge, Michael and Rosier, James L. (trans.) (1985). *Aldhelm: The Poetic Works*. Cambridge: D. S. Brewer.

Lee, N. Y. Louis, Goodwin, Geoffrey P., and Johnson-Laird, Philip N. (2008). The Psychological Puzzle of Sudoku. *Thinking & Reasoning* 14: 324–64.

Leibniz, Gottried Willhelm (1646). *Zur allgemeinen Charakteristik. Hauptschriften zur Grundlegung der Philosophie. Philosophische Werke Band 1. Übersetzt von Artur Buchenau. Durchgesehen und mit Einleitungen und Erläuterungen herausgegeben von Ernst Cassirer*. Hamburg: Verlag von Felix Meiner.

Lessing, Gotthold Ephraim (1905). *Briefe an Lessing Aus Den Jahren 1771–1773*. Berlin: Walter de Grutyer.

Lévi-Strauss, Claude (1958). *Structural Anthropology*. New York: Basic Books.

Lindgren, Harry and Frederickson, Greg (1972). *Recreational Problems in Geometric Dissections and How to Solve Them*. New York: Dover.

Lindley, Ernest H. (1897). A Study of Puzzles with Special Reference to the Psychology of Mental Adaptation. *American Journal of Psychology* 8: 431–93.

Locke, John (1690). *An Essay Concerning Humane Understanding*, P. H. Nidditch (ed.). Oxford: Clarendon Press.

Lockridge, Ross (1941). *The Labyrinth*. Westport, CT: Hyperion Press.

Loftus, Geoffrey R. and Loftus, Elizabeth F. (1983). *Mind at Play*. New York: Basic Books.

Loyd, Sam (1914). *Cyclopedia of 5000 Puzzles, Tricks and Conundrums with Answers*. New York: Dover.

Loyd, Sam (1952). *The Eighth Book of Tan*. New York: Dover.

Loyd, Sam (1959–60). *Mathematical Puzzles of Sam Loyd*, 2 vols, compiled by Martin Gardner. New York: Dover.

Lucas, Edouard A. (1882–94). *Récreations mathématiques*, 4 vols. Paris: Gauthier-Villars.

Luckiesh, Matthew (1965). *Visual Illusions*. New York: Dover.

Madigan, Jaimie (2015). *Getting Gamers: The Psychology of Video Games and Their Impact on the People Who Play Them*. Lanham, MD: Rowman and Littlefield.

Malinowski, Bronislaw (1922). *Argonauts of the Western Pacific*. New York: Dutton.

Maranda, Elli Köngäs (1976). Riddles and Riddling: An Introduction. *Journal of American Folklore* 89: 127–37.

Markov, Andrey A. (1913). An Example of Statistical Study on the Text of "Eugene Onegin" Illustrating the Linking of Events to a Chain. *Akademii Nauk, serija* VI, 3: 153–62.

Mather, Joyce E. and Kline, Linus W. (1922). The Psychology of Solving Puzzle Problems. *Pedagogical Seminary* 29: 29–282.

Matthews, William H. (1970). *Mazes & Labyrinths: Their History & Development*. New York: Dover.

Maturana, Humbero R. and Varela, Francisco (1973). *Autopoiesis and Cognition: The Realization of the Living*. Dordrecht: Reidel.

McNeill, David (1987). *Psycholinguistics: A New Approach*. New York: Harper & Row.

Merleau-Ponty, Maurice (1942). *La structure du comportement*. Paris: Presses Universitaires de France.

Metrodorus (1857–1921). *The Greek Anthology*. London: Heinemann.

Mohr, Merilyn S. (1993). *The Games Treasury*. Shelburne: Chapters.

Moscovich, Ivan (2015). *The Puzzle Universe: A History of Mathematics in 315 Puzzles*. Buffalo, NY: Firefly Books.

Müller-Lyer, Franz C. (1889). Optische Urteilstäuschunge. *Archiv für Physiologie Suppl.*: 263–70.

Netz, Reviel and Noel, William (2007). *The Archimedes Codex: Revealing the Secrets of the World's Greatest Palimpsest*. London: Weidenfeld & Nicolson.

Neuwirth, Lee (1979). The Theory of Knots. *Scientific American* 240: 110–24.

Nietzsche, Friedrich (1873). *Philosophy and Truth: Selections from Nietzsche's Notebooks of the Early 1870's*. Atlantic Heights, NJ: Humanities Press.

Northrop, Eugene (1944). *Riddles in Mathematics*. London: Penguin.

Nuessel, Frank (2011). A Note on the Names of Mathematical Problems and Puzzles. *Names* 59: 57–60.

Nuessel, Frank (2013). The Representation of Mathematics in the Media. In: *Semiotic and Cognitive Science Essays on the Nature of Mathematics*, pp. 154–98, Mariana Bockarova, Marcel Danesi and Rafael Núñez (eds.). Munich: Lincom Europa.

O'Beirne, T. H. (1965). *Puzzles and Paradoxes: Fascinating Excursions in Recreational Mathematics*. Oxford: Oxford University Press.

Ogilvie, C. Stanley (1956). *Excursions in Mathematics*. New York: Dover.

Olivastro, Dominic (1993). *Ancient Puzzles: Classic Brainteasers and Other Timeless Mathematical Games of the Last 10 Centuries*. New York: Bantam.

Ong, Walter J. (1982). *Orality and Literacy*. New York: Methuen.

Ortony, Andrew (ed.) (1979). *Metaphor and Thought*. Cambridge: Cambridge University Press.

Palmer, Adrian and Rodgers, Theodore S. (1983). Games in Language Learning. *Language Teaching* 16: 2–21.

Paulos, John A. (1980). *Mathematics and Humor*. Chicago, IL: University of Chicago Press

Paulos, John A. (1985). *I Think, Therefore I Laugh*. New York: Columbia University Press.

Peano, Giuseppe (1973). *Selected Works of Giuseppe Peano*, H. Kennedy (ed. and trans.). London: Allen & Unwin.

Pedoe, Dan (1976). *Geometry and the Visual Arts*. New York: Dover.

Peet, Thomas E. (1923). *The Rhind Papyrus*. Liverpool: University of Liverpool Press.

Peirce, Charles S. (1931–58). *Collected Papers of Charles Sanders Peirce*, ed. by C. Hartshorne, P. Weiss and A.W. Burks, Vols 1–8. Cambridge, MA: Harvard University Press.

Penrose, Lionel S. and Penrose, Roger (1958). Impossible Objects: A Special Type of Visual Illusion. *British Journal of Psychology* 49: 31–3.

Perkins, David (2000). *Archimedes' Bathtub: The Art and Logic of Breakthrough Thinking*. New York: W. W. Norton.

Perron, Paul and Danesi, Marcel (1999). *Analyzing Cultures*. Bloomington, IN: Indiana University Press.

Petkovic, Miodrag S. (2009). *Famous Puzzles of Great Mathematicians*. Providence, RI: American Mathematical Society.

Phillips, Hubert (1933). *The Playtime Omnibus*. London: Faber.

Phillips, Hubert (1934). *The Sphinx Problem Book*. London: Faber.

Phillips, Hubert (1937). *Question Time*. London: Dent.

Phillips, Hubert (1966). *Caliban's Problem Book*. New York: Dover.

Piaget, Jean (1969). *The Child's Conception of the World*. Totowa, NJ: Littlefield, Adams and Company.

Pickover, Clifford A. (2002). *The Zen of Magic Squares, Circles, and Stars*. Princeton, NJ: Princeton University Press.

Pickover, Clifford A. (2006). *The Möbius Strip: Dr. August Möbius's Marvelous Band in Mathematics, Games, Literature, Art, Technology, and Cosmology*. New York: Thunder's Mouth Press.

Plaks, Andrew H. (1996). Riddle and Enigma in Chinese Civilization. In: *Untying the Knot: On Riddles and Other Enigmatic Modes*, pp. 225–36, Galit Hasan-Rokem and David Shulman (eds.). Oxford: Oxford University Press.

Plato (2006). *Meno*, ed. by Dominic Scott. Cambridge: Cambridge University Press.

Poincaré, Henri (1902). *La science et l'hypothèse*. Paris: Flammarion.

Pollio, Howard, Barlow, Jack, Fine, Harold, and Pollio, Marilyn (1977). *Psychology and the Poetics of Growth: Figurative Language in Psychology, Psychotherapy, and Education*. Hillsdale, NY: Lawrence Erlbaum Associates.

Popper, Karl (1935). *The Logic of Scientific Discovery*. London: Routledge.

Popper, Karl (1963). *Conjectures and Refutations*. London: Routledge and Kegan Paul.

Posamentier, Alfred S. and Lehmann, Ingmar (2007). *The (Fabulous) Fibonacci Numbers*. Amherst, NY: Prometheus.

Pressman, Ian and Singmaster, David (1989). The Jealous Husbands and the Missionaries and Cannibals. *Mathematics Gazette* 73: 73–81.

Restak, Richard and Kim, Scott (2010). *The Playful Brain: The Surprising Science of How Puzzles Improve Your Mind*. New York: Riverhead.

Richards, I. A. (1936). *The Philosophy of Rhetoric*. Oxford: Oxford University Press.

Richeson, David S. (2008). *Euler's Gem: The Polyhedron Formula and the Birth of Topology*. Princeton, NJ: Princeton University Press.

Rosenhouse, Jason and Taalman, Laura (2011). *Taking Sudoku Seriously*. Oxford: Oxford University Press.

Rubin, Edgar (1915). *Synsoplevede Figurer*. Copenhagen: Nordisk Forlag.

Rubik, Ernö (1987). *Rubik's Cube Compendium*. Oxford: Oxford University Press.

Rucker, Rudy (1987). *Mind Tools: The Five Levels of Mathematical Reality*. Boston, MA: Houghton Mifflin.

Russell, Bertrand (1918). *The Philosophy of Logical Atomism*. London: Routledge.

Russell, Bertrand and Whitehead, Alfred North (1913). *Principia mathematica*. Cambridge: Cambridge University Press.

Salomon, Richard (1996). When is a Riddle not a Riddle? Some Comments on Riddling and Related Poetic Devices in Classical Sanskrit. In: *Untying the Knot: On Riddles and Other Enigmatic Modes*, pp. 168–78, Galit Hasan-Rokem and David Shulman (eds.). Oxford: Oxford University Press.

Sapir, Edward (1921). *Language*. New York: Harcourt Brace.

Saussure, Ferdinand de (1916). *Cours de linguistique générale*. Paris: Payot.

Schattschneider, Doris (1990). *M. C. Escher: Visions of Symmetry*. New York: Freeman.

Schiltz, Katelijne (2015). *Music and Riddle Culture in the Renaissance*. Cambridge: Cambridge University Press.

Schneider, Michael S. (1994). *Constructing the Universe: The Mathematical Archetypes of Nature, Art, and Science*. New York: Harper Collins.

Schuh, Fred (1968). *The Master Book of Mathematical Recreations*. New York: Dover.

Scott, Charles T. (1965). *Persian and Arabic Riddles: A Language-Centered Approach to Genre Definition*. The Hague: Mouton.

Sebo, Erin (2009). Was Symphosius an African? A Contextualizing Note on Two Textual Clues in the *Aenigmata Symphosii*. *Notes & Queries* 56: 324–6.

Selvin, Steven (1975). A Problem in Probability (letter to the editor). *American Statistician* 29: 67.

Seyeb-Gohrab, A. A. (2010). *Courtly Riddles: Enigmatic Embellishments in Early Persian Poetry*. Leiden: Leiden University Press.

Shah, Amina (1980) (trans.). *The Assemblies of Al-Hariri: Fifty Encounters with the Shaykh Abu Zayd of Serju*. London: Octagon.

Shannon, Claude E. (1948). A Mathematical Theory of Communication. *Bell Systems Technical Journal* 27: 379–423.

Shepard, Roger N. (1990). *Mind Sights: Original Visual Illusions, Ambiguities, and Other Anomalies*. New York: W. H. Freeman.

Singh, Simon (1999). *The Code Book*. New York: Anchor.

Singmaster, David (1996). Chronology of Recreational Mathematics, http://anduin.eldar.org/~problemi/singmast/recchron.html.

Singmaster, David (1998). The History of Some of Alcuin's *Propositiones*. In: *Charlemagne and His Heritage 1200 Years of Civilization and Science in Europe*, Vol. 2, pp. 11–29, P. L. Butzer, H. Th. Jongen, and W. Oberschelp (eds.). Brepols: Turnhout.

Singmaster, David (2004). Sources in Recreational Mathematics: An Annotated Bibliography, www.puzzlemuseum.com/singma/singma-index.htm.

Slocum, Jerry and Botermans, Jack (1994). *The Book of Ingenious and Diabolical Puzzles*. New York: Times Books.

Smolin, Lee (2013). *Time Reborn: From the Crisis in Physics to the Future of the Universe*. Boston, MA: Houghton Mifflin Harcourt.

Smullyan, Raymond (1997). *The Riddle of Scheherazade and Other Amazing Puzzles, Ancient and Modern*. New York: Knopf.

Spalinger, Anthony (1990). The Rhind Mathematical Papyrus as a Historical Document. *Studien zur Altägyptischen Kultur* 17: 295–337.

Sternberg, Robert J. (1985). *Beyond IQ: A Triarchic Theory of Human Intelligence*. New York: Cambridge University Press.

Sternberg, Robert J. and Davidson, Janet E. (1982). The Mind of the Puzzler. *Psychology Today* 16: 37–44.

Sternberg, Robert J. and Davidson, Janet E. (eds.) (2003). *The Psychology of Problem Solving*. New York: Cambridge University Press.

Stewart, Ian (2001). Foreword to I. Moscovich, *1000 PlayThinks*. New York: Workman Publishing.

Stewart, Ian (2008). *Taming the Infinite*. London: Quercus.

Stewart, Ian (2013). *Visions of Infinity*. New York: Basic.

Takagi, Shigeo (1999). Japanese Tangram: The Sei Shonagon Pieces. In: *The Mathemagician and Pied Puzzler: A Collection in Tribute to Martin Gardner*, pp. 97–8, E. Berlekamp and Rodgers, T. (eds.). Natick, MA: A. K. Peters.

Tan, Chuanqi, Wei, Furu, Dong, Li, Liv, Weifeng, and Zhou, Ming (2016). Solving and Generating Chinese Character Riddles. *Proceedings of the 2016 Conference on Empirical Methods in Natural Language Processing*, Austin, Texas, November 1–5, pp. 846–55. Association for Computational Linguistics.

Tarski, Alfred. (1933 [1983]). *Logic, Semantics, Metamathematics, Papers from 1923 to 1938*, John Corcoran (ed.). Indianapolis, IN: Hackett Publishing Company.

Taylor, Archer (1948). *The Literary Riddle before 1600*. Berkeley, CA: University of California Press.

Taylor, Archer (1951). *English Riddles from Oral Tradition*. Berkeley, CA: University of California Press.

Thom, René (1975). *Structural Stability and Morphogenesis: An Outline of a General Theory of Models*. Reading: Benjamin.

Trigg, Charles W. (1978). What is Recreational Mathematics? *Mathematics Magazine* 51: 18–21.

Tupper, Frederick (1903). Originals and Analogues of the Exeter Book Riddles. *Modern Language Notes* 18: 97–106.

Turing, Alan (1936). On Computable Numbers with an Application to the Entscheidungs Problem. *Proceedings of the London Mathematical Society* 42: 230–65.

Tymoczko, Thomas (1979). The Four-Color Problem and Its Philosophical Significance. *Journal of Philosophy* 24: 57–83.

Uexküll, Jakob von (1909). *Umwelt und Innenwelt der Tierre*. Berlin: Springer.

van Delft, Pieter and Botermans, Jack (1995). *Creative Puzzles of the World*. Berkeley, CA: Key Curriculum Press.

Vardi, Ilan (1998). Archimedes' Cattle Problem. *American Mathematical Monthly* 105: 305–19.

Vatuk, Prakash (1969). Amir Khusro and Indian Riddle Tradition. *Journal of American Folklore* 82: 142–54.

Verene, Donald P. (1981). *Vico's Science of Imagination*. Ithaca, NY: Cornell University Press.

Vorderman, Carol (1996). *How Math Works*. Pleasantville, NY: Reader's Digest Association.

Vygotsky, Lev S. (1961). *Thought and Language*. Cambridge, MA: MIT Press.

Wakeling, Edward (1992). *Lewis Carroll's Games and Puzzles*. New York: Dover.

Wells, David (1988). *Hidden Connections, Double Meanings: A Mathematical Exploration*. Cambridge: Cambridge University Press.

Wells, David (1992). *The Penguin Book of Curious and Interesting Puzzles*. Harmondsworth: Penguin.

Wells, David (1995). *You are a Mathematician*. Harmondsworth: Penguin.

Wheatley, Henry B. (1862). *Of Anagrams*. London: T. & W. Boone.

White, Alain C. (1913). *Sam Loyd and His Chess Problems*. Leeds: Whitehead and Miller.

Whorf, Benjamin Lee (1956). *Language, Thought, and Reality*, J. B. Carroll (ed.). Cambridge, MA: MIT Press.

Wiener, Norbert (1948). *Cybernetics, or Control and Communication in the Animal and Machine*. New York: John Wiley.

Wiener, Norbert (1950). *The Human Use of Human Beings: Cybernetics and Society*. Boston, MA: Houghton Mifflin.

Willerding, Margaret (1967). *Mathematical Concepts: A Historical Approach*. Boston, MA: Prindle, Weber & Schmidt.

Williams, Anne D. (2004). *The Jigsaw Puzzle: Piecing Together a History*. New York: Berkeley Books.

Williams, H. C., German, R. A. and Zarnke, C. R. (1965). Solution of the Cattle Problem of Archimedes. *Mathematics of Computation* 19: 671–4.

Wilson, Edward O. (1975). *Sociobiology: The New Synthesis*. Cambridge, MA: Harvard University Press.

Wilson, Edward O. (1979). *On Human Nature*. New York: Bantam.

Wilson, Edward O. (1984). *Biophilia*. Cambridge, MA: Harvard University Press.

Wilson, Edward O. and Harris, Marvin (1981). Heredity versus Culture: A Debate. In: *Anthropological Realities: Reading in the Science of Culture*, pp. 450–65, J. Guillemin (ed.),. New Brunswick, NJ: Transaction Books.

Wilson, Robin (2002). *Four Colors Suffice: How the Map Problem Was Solved*. Princeton, NJ: Princeton University Press.

Wittgenstein, Ludwig (1922). *Tractatus Logico-Philosophicus*. London: Routledge and Kegan Paul.

Wittgenstein, Ludwig (1953). *Philosophical Investigations*. New York: Macmillan.

Wundt, Wilhelm (1901). *Sprachgeschichte und Sprachpsychologie*. Leipzig: Eugelmann.

Ziolkowski, Jan. M. (1998) (ed.). *Obscenity: Social Control and Artistic Creation in the European Middle Ages*. London: Brill.

Zöllner, Johan (1860). Ueber eine neue Art von Pseudoskopie und ihre Beziehungen zu den von Plateau und Oppel beschrieben Bewegugnsphaenomenen. *Annalen der Physik* 186: 500–25.

Index

For Product Safety Concerns and Information please contact our EU representative GPSR@taylorandfrancis.com Taylor & Francis Verlag GmbH, Kaufingerstraße 24, 80331 München, Germany

Batch number: 10725973

Printed by Printforce, the Netherlands